Orelup
Aug. '99

AF779881

Reverse Osmosis and Nanofiltration

AWWA MANUAL M46

First Edition

FOUNDED
1881

American Water Works Association

MANUAL OF WATER SUPPLY PRACTICES—M46, First Edition

Reverse Osmosis and Nanofiltration

Project Manager and Technical Editor: Kathleen A. Faller
Copy Editor: Phillip Murray
Production Editor: Alan Livingston

Library of Congress Cataloging-in-Publication Data

Simplified procedures for water examination. -- 4th ed.
xvii, 223 p. 17×25 cm. -- (AWWA manual ; M46)
Includes bibliographical references and index.
ISBN 0-89867-978-8
1. Water--Purification--Reverse osmosis process. 2. Water --Purification--Membrane Filtration. 3. Drinking water --Purification. I. Series.
TD491.A49 No. 46
628.1 s--dc21 97-25864
[628.1'64] CIP

Printed in the United States of America
American Water Works Association
6666 West Quincy Avenue
Denver, CO 80235

ISBN 0-89867-978-8

Printed on recycled paper

Contents

Figures

This page intentionally blank.

Tables

Preface

Reverse osmosis (RO) and nanofiltration (NF) are membrane technologies that can be applied to the treatment of various water sources for the production of drinking water. Membrane technologies can remove organic and inorganic substances from water and can replace or be adjuncts to such traditional treatment methods as sand filtration, primary disinfection, lime and soda softening, ion exchange, and evaporative processes. Reverse osmosis and NF are both pressure-driven membrane processes with similar process configurations and equipment. The main differences between the processes are primary treatment focus (demineralization or salt reduction for RO and hardness and/or organics removal for NF) and the degree of necessary applied pressure (higher for RO by a factor of two or more). However, the two membrane processes exhibit far more similarities than differences, thus their inclusion in the same manual.

Membrane technology became particularly applicable to drinking water production in the late 1980s when the amendments to the US Environmental Protection Agency's Safe Drinking Water Act (SDWA) required higher-quality drinking water. The amendments were based on health effects research and the ability to detect contaminants at increasingly lower detection limits using sensitive analytical techniques. Because of their ability to remove or reduce many of the substances addressed by the SDWA, membranes will play an increasingly important role in enabling water utilities to meet these regulations.

This manual was developed to provide an overview of RO and NF technology for operators, administrators, engineers, scientists, educators, and anyone seeking an introduction to these processes. Chapters on theory and applications, design, equipment and installation, and operations and maintenance are presented. The information contained in these chapters includes history, regulations, theory, terminology, water resources, pretreatment, process technology, posttreatment, concentrate disposal, membrane manufacturing, configuration, maintenance, chemistry, costing, and safety.

As this is the first edition of AWWA Manual M46, *Reverse Osmosis and Nanofiltration,* the Membrane Processes Committee and the American Water Works Association welcome comments and suggestions for improving future editions. Please contact the Water Quality Engineer at AWWA, (303) 794-7711, 6666 W. Quincy Ave., Denver, CO 80235.

Acknowledgments

This new AWWA manual originated with, and was written and approved by, the Membrane Processes Committee (819), Water Quality Division, initially under the direction of past chair, William J. Conlon. At the time of approval, the review committee consisted of the following:

S.J. Duranceau, Chair, Boyle Engineering, Orlando, Fla.
J.Q. Adams, US Environmental Protection Agency, Cincinnati, Ohio
G.L. Amy, University of Colorado, Boulder, Colo.
R.P. Carnahan, University of South Florida, Tampa, Fla.
W.J. Conlon, former Chair, Rust Environment & Infrastructure, Sheboygan, Wis.
P. Côté, Compagnie Générale des Eaux, Paris, France
E.P. Geishecker, Ionics Inc., Watertown, Mass.
E.A. Ibrahim, American Water Works Service Company, Voorhees, N.J.
H.S. Kohli, Fluid Systems Corporation, San Diego, Calif.
M. LeRoy, Florida Department of Environmental Protection, Tallahassee, Fla.
J.C. Lozier, CH2M Hill, Phoenix, Ariz.
S.A. McClellan, The Dow Chemical Company, West Palm Beach, Fla.
R.O. Myers Jr., DuPont Permasep Products, Newark, Del.
R.K. Noack, HDR Engineering Inc., Austin, Texas
D.L. Rohe, Montgomery Watson, Pasadena, Calif.
W.B. Suratt, Camp Dresser & McKee, Vero Beach, Fla.
M.A. Thompson, Malcolm Pirnie Inc., Yorktown, Va.
R.T. Wale, Memtec America Corporation, Timonium, Md.
R.A. Bergman, CH2M Hill, Gainesville, Fla.
J.G. Jacangelo, Montgomery Watson, Herndon, Va.
J.S. Taylor, University of Central Florida, Tampa, Fla.
D.M. Thompson, Camp Dresser and McKee, Jacksonville, Fla.

Authors of the manual represent academic, consulting, manufacturing, and utilities backgrounds and include the following:

William J. Conlon, Rust Environment & Infrastructure, Sheboygan, Wis.
Steven J. Duranceau, Chair, Boyle Engineering, Orlando, Fla.
Harbans S. Kohli, Fluid Systems Corporation, San Diego, Calif.
James C. Lozier, CH2M Hill, Phoenix, Ariz.
Irving Moch Jr., DuPont Permasep Products, Newark, Del.
Stuart A. McClellan, The Dow Chemical Company, West Palm Beach, Fla.
Thomas M. Missimer, Missimer International, Fort Myers, Fla.
Oram J. Morin, Black & Veatch, Orlando, Fla.
David H. Paul, David H. Paul Inc., Farmington, N.M.
Hermann Pohland, DuPont Permasep Products, Newark, Del.
Larry R. Reitz, Pfizer Inc., Groton, Conn.
James S. Taylor, University of Central Florida, Orlando, Fla.
Donald M. Thompson, Camp Dresser and McKee, Jacksonville, Fla.

Photographs used in the manual were provided by Boyle Engineering Corporation; Gulf Environmental Systems; PWT Americas, Inc.; and Water Equipment Technologies, Inc.

Chapter 1

Introduction

This chapter provides an overview of membrane processes, presents appropriate applications in the treatment of water for potable uses, and introduces the main theoretical concepts. Most of the terms used in the manual are introduced in this chapter; they are also defined in the glossary at the end of the manual. For those unfamiliar with the general aspects of a membrane plant, Figure 1-1 shows a 1.9-mgd, low-pressure reverse osmosis (RO) plant (Acme Improvement District in Wellington, Fla.). As can be seen from the photograph, membrane facilities can provide significant treatment capacity in limited space.

OVERVIEW

Five membrane processes commonly used in the production of drinking water are RO, nanofiltration (NF), ultrafiltration (UF), microfiltration (MF), and electrodialysis/electrodialysis reversal (ED/EDR). Significant differences exist in the mechanisms of contaminant removal for drinking water treatment applications among these five membrane processes. Other membrane processes, such as gel filtration and two-film gas membranes, are presently in research and development and are not covered in this manual. Nanofiltration and RO are pressure-driven and diffusion-controlled membrane processes. Microfiltration and UF are also pressure-driven membrane processes; however, removal of dissolved substances is based on a sieving mechanism. These two processes do not remove ions. Microfiltration and UF, like RO and NF, can remove turbidity and microorganisms. Electrodialysis/electrodialysis reversal are charge-driven membrane processes that remove ions, but not turbidity or microorganisms, from feedwater.

Development of Membrane Processes

At the University of California at Los Angeles in the early 1960s, Loeb developed a membrane to use in producing drinking water from seawater (Weber 1972). This relatively thick membrane was made from cellulose acetate. Seawater was forced against the membrane at high pressure (greater than 1,000 psi).

Figure 1-1 Photograph of a 1.9-mgd, low-pressure reverse osmosis plant

Cellulose acetate membranes have relatively tight, nonporous surfaces; that is, they have no holes. Under high pressure, water moves through the molecular structure of the membrane surface via a process called convection. The dissolved solids (solutes) also pass through the membrane in a process called diffusion, but at a much lower rate than water. Consequently, the water (permeate) that passes through the membrane contains fewer dissolved solids than does water entering the system (feedwater). The amount of energy (hydraulic pressure) required to drive feedwater through the membrane depends on the membrane's thickness and tightness and the osmotic pressure of the feedwater. Osmotic pressure is the pressure required to prevent the movement of water from a dilute solution (lower dissolved solids concentration) across a semipermeable membrane to a more concentrated solution (higher dissolved solids concentration). Thus, energy or hydraulic pressure is required to overcome both the physical resistance of the membrane and the chemical resistance of the more concentrated feedwater stream.

Applications for membranes increased as less saline, brackish waters needed to be treated, requiring the development of new membrane materials. Such applications include membrane processes for treating seawater (high-pressure RO, 800–1,200 psi), brackish water (low-pressure RO, 250–400 psi), and fresh water (NF, 75–150 psi, also called *membrane softening*). As the efficiency of membrane applications has increased, resulting in lower operating pressures, the energy costs to drive the processes have decreased.

Both RO and NF membranes are manufactured using organic polymers in several physical configurations. The two primary materials used to manufacture RO and NF membranes include cellulose acetate and its derivatives and various polyamides, often referred to as *thin-film composites*. Microfiltration and UF membranes are also made from cellulose acetate and polyamide materials as well as other organic polymers, such as polyacrylonitrile, polysulphone, and polyethersulphone in the form of sheets, woven tubes, or hollow fibers. The major configurations for RO or NF membranes are spiral-wound and hollow fiber. Configuration, materials, construction, and flow patterns are discussed throughout this manual.

Research and practice have proven membrane processes to be reliable and consistent methods of removing contaminants from drinking water. Reverse osmosis and ED/EDR have been used primarily to desalt brackish waters. Reverse osmosis is used extensively to desalt seawater for potable use. Nanofiltration, developed in the late 1970s, is applicable for trihalomethane (THM) precursor rejection and for softening of groundwaters normally treated by lime softening. Ultrafiltration has been used extensively in industry to separate high-molecular-weight organic molecules from solutions. Microfiltration also is used in industry, primarily to remove fine particles or turbidity. The pressure-driven processes of RO, NF, and UF are capable of extensive, if not complete, pathogen removal. Microfiltration is also capable of substantial pathogen removal, although it is also possible that significant pathogens can pass through the membrane before the MF cake layer builds up.

In summary, membrane processes are capable of desalting, softening, and removal of THM precursors, pathogens, suspended particulate material (as indicated by turbidity measurements), and specific organic compounds.

Figure 1-2 presents an overview of membrane and conventional filtration processes. Table 1-1 also summarizes the same information in tabular form.

Reverse Osmosis and Nanofiltration Membrane Systems

A typical RO or NF membrane system consists of three separate subsystems: pretreatment, the membrane process, and posttreatment. There is one input (a feed stream) and two outputs (a permeate stream and a concentrate stream). Figure 1-3 illustrates the typical system.

Essentially all sources of water used as feedwater for membrane systems must undergo certain levels of pretreatment. Typical pretreatment processes usually involve adding acid, antiscalant, or both (to prevent precipitation of salts during membrane filtration) followed by 5- to 15-μm cartridge filtration. Additional pretreatment methods (not shown in Figure 1-3) usually occur before typical pretreatment and are necessary when the feedwater contains excessive fouling materials. After pretreatment, water passes through the membrane array, which consists of active RO or NF membrane surfaces with effective pore sizes of about 300 to 10,000 μm. Posttreatment includes many unit operations common to drinking water treatment, such as aeration–degasification, disinfection, and corrosion control. The unit operations in each subsystem will be discussed in detail later in this manual.

A thorough understanding of source water characteristics is essential to the successful design and operation of a membrane plant. The contaminants found in the source water determine which combination of treatment techniques will be effective. As is explained in chapter 2, few, if any, RO or NF membrane plants use surface water sources. However, US Environmental Protection Agency (USEPA)

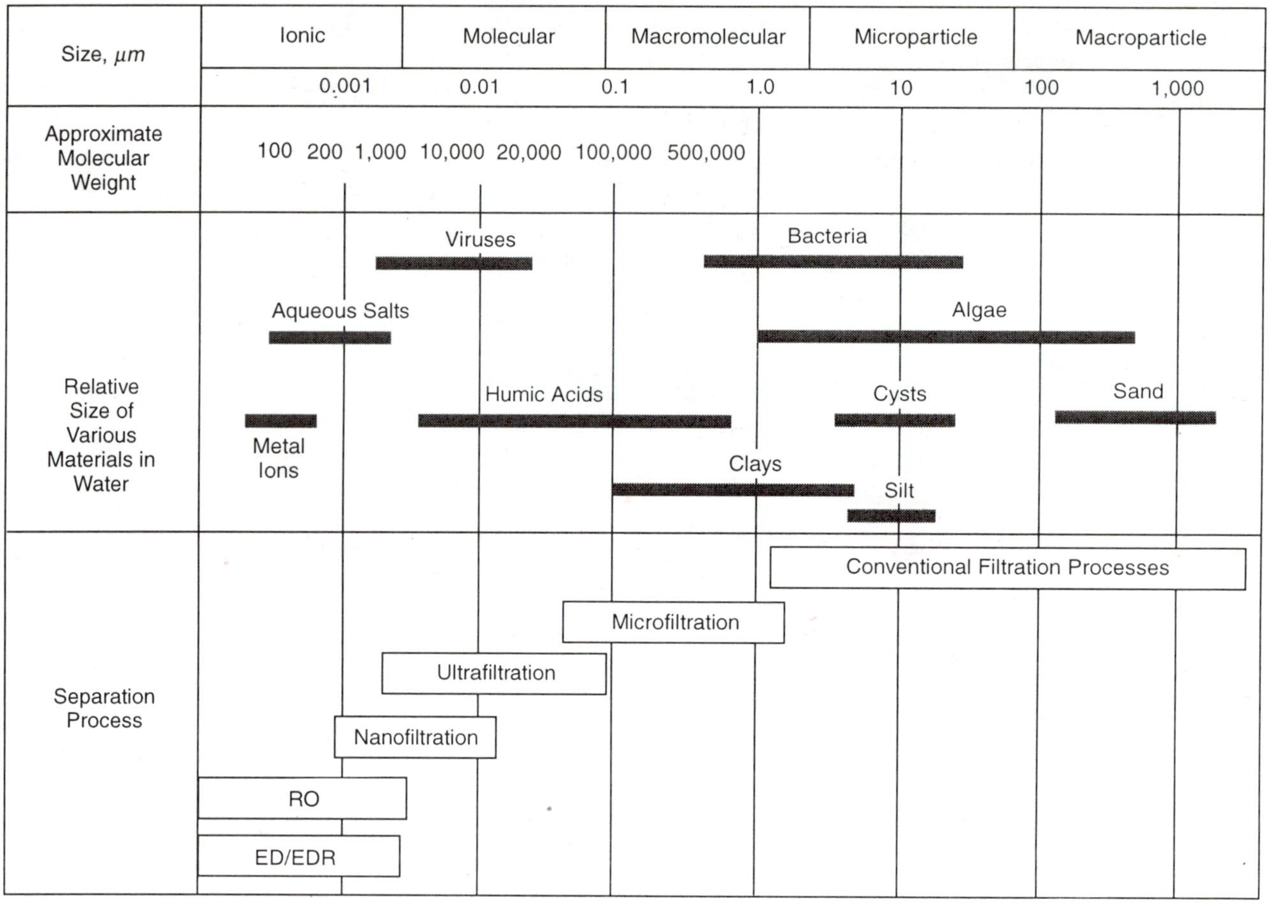

Figure 1-2 Membrane and conventional process overview

Table 1-1 Drinking water membrane separation technologies and their traditional counterparts

Membrane Separation Technology	Substances Removed	Comparable Traditional Water Treatment Methods
Microfiltration	Bacteria and larger colloids; separation of precipitates and coagulates	Ozonation–ultraviolet radiation, chlorination, sand filters, bioreactors, and coagulation-settling tanks
Ultrafiltration	All of the above, plus viruses, high-molecular-weight proteins, organics, and pyrogen	Sand filters, bioreactors, and activated carbon
Nanofiltration	All of the above, plus divalent ions, larger monovalent ions, color, and odor	Lime–soda softening and ion exchange
Reverse osmosis	All of the above, plus monovalent ions	Distillation, evaporation, and ion exchange
Electrodialysis/ electrodialysis reversal	Dissolved ionic salts	Ion exchange

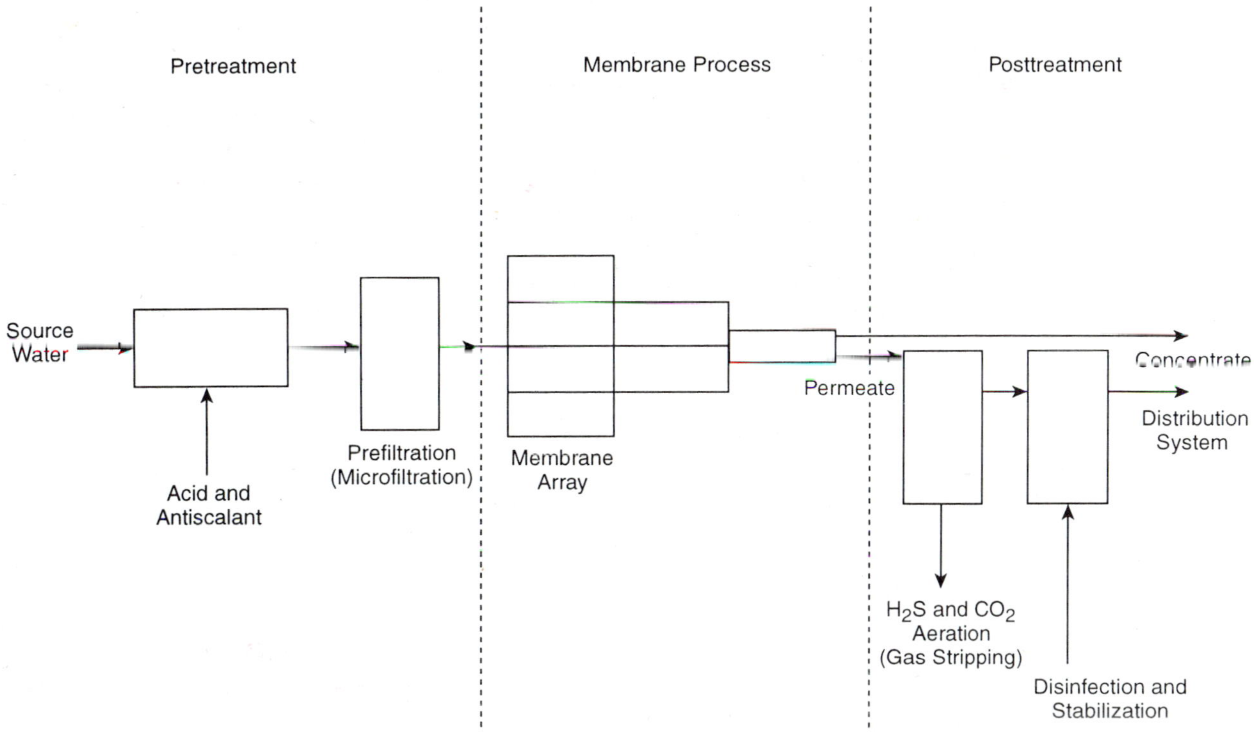

Figure 1-3 Typical reverse osmosis or nanofiltration membrane system

regulations will require all water treatment plants with total organic carbon (TOC) greater than 4 mg/L to evaluate membrane or granular activated carbon (GAC) technology; this will increase the use of membranes for treating surface water.

Concentrate from RO and NF membrane processes is regulated as an industrial waste in some states; it often requires significant treatment before disposal. Permits for concentrate disposal must be obtained from the USEPA, a state regulatory agency, or both (Mickley et al. 1994).

Existing Reverse Osmosis and Nanofiltration Plant Performance

The following paragraphs review the results of a survey of the operation and reported finished water quality of RO and NF plants (Taylor et al. 1989a). A survey form was mailed to 117 membrane plants, including all potable water membrane plants that were identified in the United States. Twenty brackish water RO or NF plants responded, providing insight into the operation and performance of membrane plants. All responding facilities used groundwater sources, and no seawater RO plants responded to the survey.

The most common RO system used the following processes, in sequence, antiscalant addition (sometimes in combination with sulfuric acid), prefiltration, membranes, aeration–degasification, chlorine disinfection, and sometimes stabilization with sodium hydroxide. Reported RO plant capacities ranged from 800 gpd to 14 mgd, with an average capacity of 0.5 mgd. The average operation and maintenance cost reported was $1.28 per 1,000 gal of permeate produced. This cost was divided into management (22 percent), power (33 percent), chemical (16 percent), maintenance (7 percent), and other costs (22 percent), including membrane replacement. The average ratio of assigned personnel per 1 mgd of capacity was 3.2.

Reported membrane-cleaning frequencies typically ranged from every two months to every six months. Spiral-wound, thin-film composite, or hollow fiber cellulose acetate membranes were used most frequently. Average feed pressure for the respondents was 290 psi. The reported methods of concentrate disposal and the number of plants using each method were as follows: saline water discharge (10), land spreading (4), sewer discharge (2), deep well injection (1), and infiltration basin (1). All plants were required to obtain a National Pollutant Discharge Elimination System permit for concentrate disposal.

Except for one NF plant, all responding plants were designed to remove total dissolved solids (TDS) from brackish water. Product recovery ranged from 38 to 85 percent and averaged 70 percent. The reported finished water quality of the responding RO plants averaged 209 mg/L TDS and pH 7.3. The average total hardness and alkalinity were 32 and 54 mg/L as calcium carbonate, respectively. Only 4 of the 20 responding plants reported THMs, which averaged 33 μg/L. Finished water turbidity averaged 0.85 nephelometric turbidity units (ntu), a relatively high value. (This turbidity may have been formed following membrane filtration by the precipitation of sulfur during aeration for the removal of hydrogen sulfide). On average the permeate water was poorly buffered and corrosive, and none of the RO plants practiced alkalinity recovery. The average source water dissolved organic carbon and color of the 13 sampled RO plants were 1.8 mg/L as *C* and 11 chloroplatinate units (cpu), respectively, which were reduced to 0.3 mg/L and 2 cpu in the finished water.

The study also documented, through US state regulatory agencies, that over 100 membrane plants were consistently producing water at an acceptable cost and quality. Reverse osmosis and NF plants have been producing hundreds of millions of gallons of drinking water every day for the last decade and appear to offer some of the best treatment techniques for meeting drinking water quality regulations.

REVERSE OSMOSIS AND NANOFILTRATION MEMBRANE APPLICATIONS

This section presents the current and developing applications of RO and NF membrane processes. Figure 1-4 is a flowchart for selecting the appropriate membrane process and can serve as a summary of the following discussion.

Desalting (Total Dissolved Solids Removal)

One of the first applications for membrane technology was the conversion of seawater into drinking water through RO. Early cellulose acetate membranes were much less permeable than current membranes. The disadvantages of early membranes included the high pressure involved (over 1,000 psi) and the low recovery rate (only 10 to 25 percent of the source water was recovered as product). Power costs alone exceeded $4.00 per 1,000 gal of permeate.

The first commercial application of RO membranes for brackish water desalting began in the early 1960s. This application used the spiral-wound configuration developed in 1967 by General Atomics. E.I. DuPont de Nemours, Inc. (DuPont) introduced the polyamide hollow fine fiber in 1969 in the form of the B-9 permeator for brackish water desalting. These brackish water modules generally operated in the pressure range of 300–400 psi. The first municipal brackish water RO plant was located at Florida's Ocean Reef Club. The plant began operations in October 1971

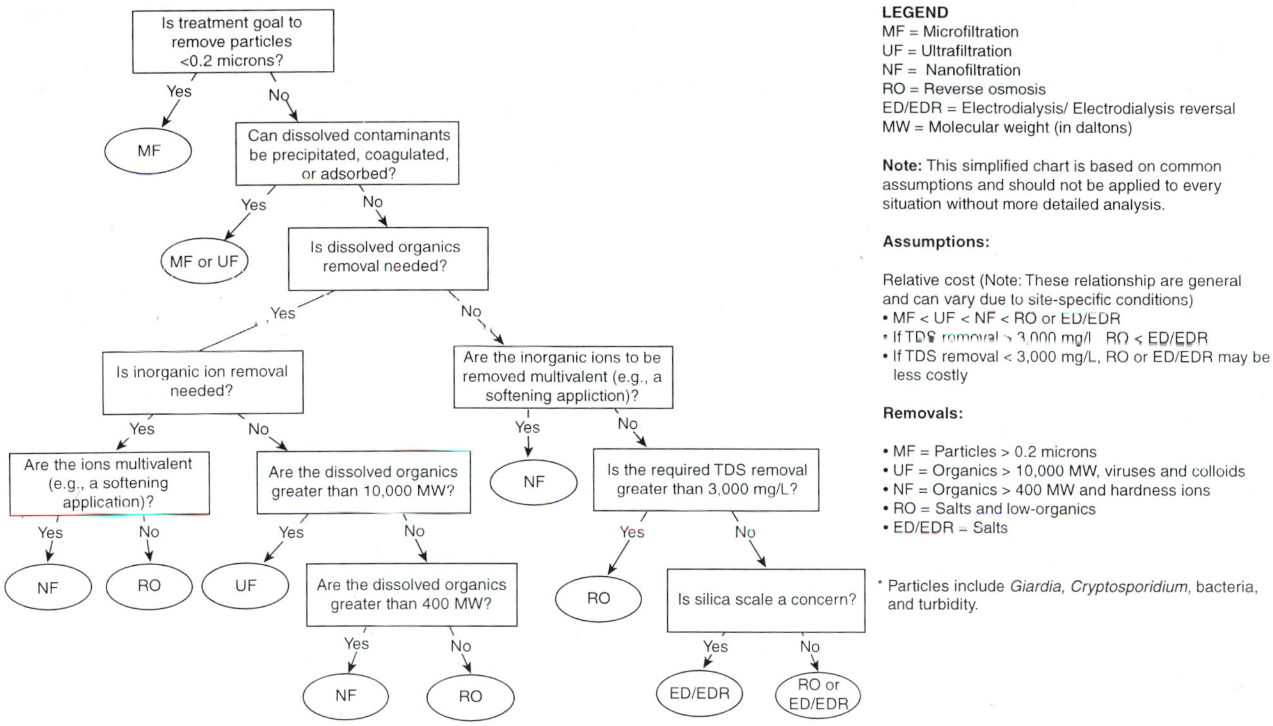

Figure 1-4 Generalized membrane process selection chart

with an initial operating pressure of 600 psi, and a 0.6-mgd capacity, which was later expanded to 0.93 mgd.

In 1974, DuPont introduced the B-10 permeator, the first RO membrane capable of producing potable water from typical seawater in one membrane pass at operating pressures of 800–1,000 psi. Spiral-wound, thin-film composite RO membranes developed for both seawater and brackish water desalting were introduced in the mid- to late 1970s. Feed pressures for the early composite membranes were approximately the same as for the cellulosic and polyamide hollow fiber modules.

Dow Chemical Company's introduction of the low-pressure Dowex™ hollow fiber RO membrane led to a major reduction in the cost of brackish water RO operation. The first plant to use the new membrane began operation in 1981 at Venice, Fla., with a 1-mgd capacity. The Dowex™ membrane, which is no longer in production, provided salt rejection (SR) and flux rates comparable to the standard-pressure cellulosic and polyamide membranes at roughly one half the operating pressure (200–250 psi, as compared to 400–600 psi).

Low-pressure, thin-film composite, spiral-wound modules were first introduced in the early 1980s by Filmtec Corporation and Fluid Systems. These membranes are now commonly used, except where the chlorine tolerance of cellulosic membranes is desired or where high-salinity (high-osmotic-pressure) feedwaters require standard and high-pressure composite membranes with greater SR capacity.

The expansion of the Englewood Water District's RO plant in southwestern Florida illustrates the evolution of spiral-wound brackish water RO membrane technology. The original (1982) RO process trains used standard brackish water cellulose acetate blend membranes operating at 400–600 psi. Reverse osmosis trains installed during an initial expansion in 1986 employed the early-generation polyurea composite membranes. During a 1989 expansion, the new trains and several of the older units were outfitted

Table 1-2 Comparison of three generations of brackish water reverse osmosis membranes at Englewood Water District

	RO Train			
Parameter	Train A (1982), Cellulose Acetate Blend	Train A (1986), Composite (Polyurea)	Trains B and C (1986), Composite (Polyurea)	Trains D and E (1989), Composite (Fully Aromatic Polyamide)
Design feedwater TDS, *mg/L*	4,700	6,000	6,000	7,000
Maximum design feed pressure, *psi* (stage 1)	600	420	*	335
Normalized specific productivity, *gpd/psi*	0.041	0.057	0.069	0.082
Relative membrane energy usage, % of initial plant design	100	72	59	50

*Trains include interstage booster pumps between the first and second stages.

NOTE: All RO trains have the following characteristics: 0.5-mgd permeate capacity at 65 percent minimum and 7 percent design recovery. All trains are normalized to 4,700 mg/L feedwater TDS to permit direct comparison of energ usage atinitial plant feedwater condition

with the most advanced, low-pressure, fully aromatic polyamide composite membranes. As illustrated in Table 1-2, plant energy for membrane operation today is 50 percent of the amount used in the original plant design.

Reverse osmosis is currently the most commonly used membrane process for removing TDS in drinking water applications. Process power costs have been reduced to significantly less than $1.00 per 1,000 gal of permeate and, in many cases, to less than $0.50 per 1,000 gal of permeate. More than 100 drinking water plants in the United States use RO to produce drinking water from brackish water. Although few desalination processes treat seawater in the United States, several major facilities in Middle Eastern countries use RO for seawater desalination.

Disinfection By-Product Precursor Removal

Disinfection by-product (DBP) precursors can be significantly reduced by RO or NF membrane treatment. Disinfection by-product precursors form when natural organic matter reacts with chlorine or other chemical oxidants used or produced in the disinfection process. Disinfection by-product precursors include total trihalomethanes (TTHMs) and haloacetic acids (HAAs). In the United States, TTHMs are regulated at 100 µg/L and are scheduled to be regulated at lower maximum contaminant levels (MCLs) by the Disinfectant–Disinfection By-product (D–DBP) Rule. Haloacetic acids are not currently regulated. Phase 1 D–DBP Rule MCLs are anticipated to be 80 µg/L TTHMs and 60 µg/L HAAs. Phase 2 D–DBP Rule MCLs may be regulated at 40 µg/L TTHMs and 30 µg/L HAAs.

In the United States, the Information Collection Rule requires that large utilities with TOC levels greater than 4 mg/L must investigate either GAC or membranes to reduce TOC levels to less than 4 mg/L. Reverse osmosis or NF alone has been shown in many cases to reduce TOC to less than 0.5 mg/L, TTHMs to less than 30 µg/L, and HAAs to less than 20 µg/L, depending on the concentration of DBP precursors in the feedwater. Reverse osmosis and NF membranes' molecular weight cutoffs of 500 or less have been found to reject more than 90 percent of DBP precursors from highly organic surface waters and groundwaters (Taylor et al. 1989b).

Hardness, Color, and Turbidity

In 1976, Paul Culler and Stuart McClellan introduced the concept of membrane softening. The following year a 0.25-mgd membrane softening plant was installed in Pelican Bay, Naples, Fla. The plant used a Fluid Systems hydrolyzed, or *loose*, RO membrane and was later expanded to 0.5 mgd. Membrane softening saw limited use until 1986 when Filmtec Corporation (now part of Dow Chemical Company) introduced NF membranes. Other manufacturers subsequently developed similar membranes. Nanofiltration became a significant alternative to lime softening for reducing the level of calcium and magnesium ions in naturally hard waters where TDS reduction is not a primary treatment goal. Nanofiltration is also more effective than lime softening in removing naturally occurring color, DBP precursors, and turbidity.

It is expected that turbidity will be regulated by the Enhanced Surface Water Treatment Rule as a primary standard, with color regulated as a secondary standard. Hardness will not be regulated but is often a targeted constituent, as in a softening process. Typically, NF removes 60 to 80 percent of total hardness, more than 90 percent of color, and virtually all turbidity in the feedwater. However, NF should not be used for turbidity removal because the particles producing turbidity, if not removed by pretreatment, will plug or foul the membrane, making the process uneconomical. Figure 1-5 shows a photograph of a 12-mgd membrane softening plant in Plantation, Fla.

Figure 1-5 Photograph of a 12-mgd membrane softening plant

Inorganic Chemical Removal

The USEPA currently recognizes RO as the best available technology for removing most inorganic chemicals (IOCs) regulated under the Safe Drinking Water Act (Clark and Parrotta 1991). This classification reflects the broad-spectrum removal capability of the RO process. Reverse osmosis is also being proposed as a BAT for other IOCs currently being considered for future regulation, including radionuclides, arsenic, sulfate, and boron.

The ability of NF to remove regulated heavy metal IOCs is determined to a large extent by the specific dissolved solids character of the water. To a greater degree than for the RO process, NF's rejection of specific multivalent cations is a function of solution pH and the types of co-ions present. Actual testing is necessary to make an accurate prediction of percent removals. Based on studies conducted for hardness removal, NF appears to be applicable for removal of certain IOCs within limits.

Specifically, RO and NF systems can be designed to have appropriate post membrane treatment unit processes to produce a noncorrosive water and to meet the MCLs for the Lead and Copper Rule, which requires copper and lead concentrations to be less than 1.3 mg/L and 0.15 µg/L respectively, for 90 percent of the required number of samples in their service area. Copper and lead corrosion is complex and depends on many different water quality parameters, including alkalinity and pH values. Both RO and NF systems can be designed to produce a well-buffered, low-TDS water that is easily stabilized by downstream neutralization. Buffering the permeate for RO and NF systems is described in the posttreatment section of this chapter.

Nitrate Removal

Nitrate is regulated in public drinking waters because it can interfere with oxygen uptake by blood hemoglobin once it has been converted to nitrite. Reverse osmosis is considered an effective nitrate removal process for groundwater supplies polluted by the agricultural use of nitrate-containing fertilizers. Rejection of nitrate by some RO membranes is significant; composite polyamide low-pressure brackish water membranes typically exhibit sodium nitrate rejection in the range of 93–97 percent (Filmtec 1988; Toray 1989).

Several RO plants are currently in operation in southern California treating groundwaters contaminated with high concentrations of nitrate from past agricultural practices, including those in the cities of Riverside and Tustin. Total dissolved solids levels of the nitrate-containing groundwaters at these locations also exceed the recommended drinking water level, making RO more cost-effective than ion exchange for such applications. The successful operation of these plants has led to their expansion or the future development of additional facilities at other locations. A nitrate removal plant using RO is in operation in Brighton, Colo. (Brunswick, Suratt, and Burke 1993). This application of RO processes should become more common as more municipalities attempt to reclaim nitrate-laden groundwaters.

Fluoride Removal

Many groundwaters in the United States contain excess levels of naturally occurring fluoride. One such region is coastal South Carolina, where groundwater from the Black Creek aquifer supplies several municipalities. Fluoride levels in the aquifer range from 2 to nearly 6 mg/L, as compared to the water standards for fluoride of 4 mg/L to protect against skeletal fluorosis, and 2 mg/L to protect against dental fluorosis. Total dissolved solids levels also typically range from 600 to 1,200 mg/L, a range greater than desirable levels. The town of Mount

Pleasant, S.C., has installed three wellhead treatment plants using RO to reduce ground-water fluoride levels from as high as 6 mg/L down to 1.0 mg/L. The Mount Pleasant RO plant also reduces TDS. Wellhead RO plants for fluoride control have been implemented in other municipalities in the coastal South Carolina area. In most cases, treatment costs can be reduced through the use of split treatment, in which a portion of adequately treated source water is bypassed around the membrane system and blended with the RO permeate. Pilot testing on a freshwater aquifer with low TDS at Twentynine Palms, Calif., has shown a reduction in fluoride from 7 mg/L to less than 0.2 mg/L in the permeate.

Removal of Synthetic Organic Chemicals

Synthetic organic chemicals (SOCs) are commonly found as pesticide residuals. Pilot testing has been performed by a municipality in association with the USEPA to evaluate a number of different types of RO membranes for pesticide removal from a groundwater contaminated by agricultural chemicals (Bailer et al. 1987). The study found that removals were greatest for the polyamide thin-film composite membranes (67–95 percent), and it concluded that RO should be seriously considered as a water treatment process for this application (Lykins et al. 1988).

Nanofiltration is receiving increased attention for use in removing pesticides from surface waters in Europe as part of a more comprehensive water quality improvement goal. Studies in the Netherlands and France have assessed the capability of a wide range of NF membranes to remove commonly occurring compounds such as simazine and others to below the 0.1 μg/L limit for total pesticides (Hofman et al. 1993, Côté et al. 1993). Results indicate that NF can meet treatment goals but that removals are highly dependent on membrane material. Nanofiltration will likely become an integral process for future surface water treatment plants in these countries.

Because of the relatively recent introduction of NF, research on its potential for SOC removal is limited. Theoretically, specific SOC rejection would be primarily a function of molecular size and degree of ionization. This theory is corroborated by a pilot study demonstrating that the degree of rejection is proportional to the molecular weight. Synthetic organic chemicals with a molecular weight greater than 300 were completely rejected by one type of NF membrane, while those with molecular weights less than 300 were only partially rejected. For these studies, the degree of rejection was proportional to the molecular weight (Taylor et al. 1989b).

Pathogen Removal

Reverse osmosis and NF membranes provide an effective barrier to the passage of protozoa, bacteria, and viruses. However, RO and NF systems are not normally considered for pathogen removal alone because they have relatively higher construction and operating costs when compared with conventional treatment, MF, and UF.

Still, this capacity for pathogen removal enhances the applicability of RO and NF for treating surface water (and groundwater under the direct influence of surface water) in cases when these technologies are being considered to meet other water quality goals as well (such as desalting or IOC or DBP precursor removal). Few studies have been conducted to provide empirical evidence of the log-removal capability of RO and NF for pathogens. Consequently, health regulatory agencies have been reluctant to grant log-removal credits greater than those normally afforded by conventional treatment. One such study has demonstrated that RO membranes provide between 4- and 5-log removal of viruses normally associated with waterborne disease (Lozier, McKim, and Rose 1994). Additional studies are needed to clearly define pathogen removals by RO and NF under controlled and field studies, as well as to determine the degree of risk associated with leakage resulting from physical damage to the membrane or other mechanical components of the membrane assembly.

Indirect Potable Reuse

Reverse osmosis and NF are being increasingly used in the reclamation of municipal wastewaters serving indirectly as future potable water supplies. In this application, RO and NF remove a host of contaminants, including nitrogen, heavy metals, TOC, and pathogens. The most widely known and oldest facility of this type is Water Factory 21 in Fountain Valley, Calif., which produces a high-quality reclaimed water for direct injection into a potable water aquifer (groundwater recharge). This facility is scheduled for expansion, and several other municipalities are currently investigating groundwater recharge using RO or NF in both California and Arizona. San Diego, Calif., is evaluating the reuse of wastewater to supplement surface potable supplies, using RO as an integral treatment technology. Indirect potable reuse of domestic wastewater represents a major future application for RO and NF.

MEMBRANE PROCESS THEORY

Many types of membranes and processes with different performance characteristics are available. Selection of a specific membrane depends on

- the process the membrane uses to separate the components of the feedwater (e.g., RO and ED/EDR membranes use different processes)
- the efficiency of the membrane process in separating the feedwater components
- the stability of the membrane, as well as the compatibility with the feedwater components to be separated

In general, feedwater component separation is accomplished based on one of the following:

- molecular size differences between the components and the membrane polymer structure, as in MF, UF, NF, or RO
- electrical charge as in ED/EDR
- solubility of the different components in the membrane phase, as in dialysis

The RO process can remove more than 99 percent of all dissolved minerals and organic compounds, as well as biological and colloidal suspended matter, including turbidity, from water. The NF process is primarily used for organic removal and water softening, and it can remove 5 to 95 percent of TDS.

In any membrane process, flow characteristics are unique functions of the membrane polymer. For example, product flow or solute (salt) passage for a polyamide polymer is different from that of a cellulose acetate polymer, which is different from that of a polysulfone polymer. If the membrane area in a device is twice that of another device using the same polymer, the water flow will be double. Similarly, a thicker membrane will give less water flow than a thinner membrane of the same polymer.

The pressure applied in the system and the concentration differential across the membrane can be adjusted by an end user. Other factors, such as the membrane material, membrane area in a device, and membrane thickness, are set and controlled only by the membrane supplier. The following paragraphs discuss the adjustable variables for RO and NF.

In natural osmosis, a semipermeable membrane separates pure water (solvent) from a salt (solute) solution, as shown in Figure 1-6. Pure water naturally passes through the membrane to dilute the salt solution; the driving force for the flow is the difference in chemical potential between the two solutions. Chemical potential is directly proportional to changes in temperature and pressure and inversely proportional to changes in solute concentration. Water flow continues until the pressure created by the hydraulic head equals the osmotic pressure of the salt solution. At this point the two liquids are said to be at osmotic equilibrium (Figure 1-7).

In RO, applying external pressure on the salt solution allows the natural flow of water to be reversed (Figure 1-8), hence the term *reverse osmosis.* Nanofiltration is a similar process to RO, but NF's "looser" membranes require a lower driving force to move water through the membrane.

Design Equations

This section describes the fundamentals associated with diffusion-controlled membrane processes, such as RO and NF. In such processes, water is forced through the membrane by a pressure differential (a type of motion called *convection*), and dissolved salts pass through the membrane because of a concentration differential (a type of motion called *diffusion*). Because the membrane is semipermeable, both dissolved salts and water are transported to differing degrees through the molecular structure of the active surface layer of the membrane. The osmotic pressure for a salt solution can be estimated as 1 psi per 100 mg/L TDS for fresh and brackish waters. At saline salt concentrations (i.e., 35,000 mg/L TDS) the ratio of pounds per square inch to milligrams per litre TDS is greater than 1.0 (Weber 1972).

The standard model for describing the RO and NF processes is the homogenous solution diffusion (HSD) model, which illustrates the effects of feed concentration, membrane characteristics, recovery, and pressure on permeate concentration. The HSD model incorporates operational parameters (water quality, water recovery, and membrane-specific coefficients) to determine system performance. The HSD model does not consider the effects of certain chemical and physical constraints, and it makes

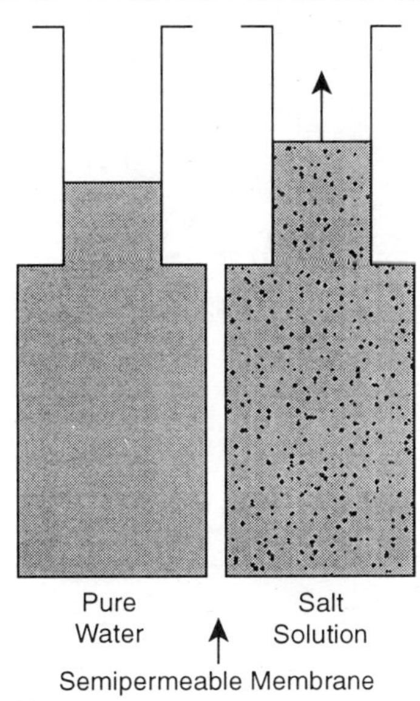

Figure 1-6 Osmotic flow

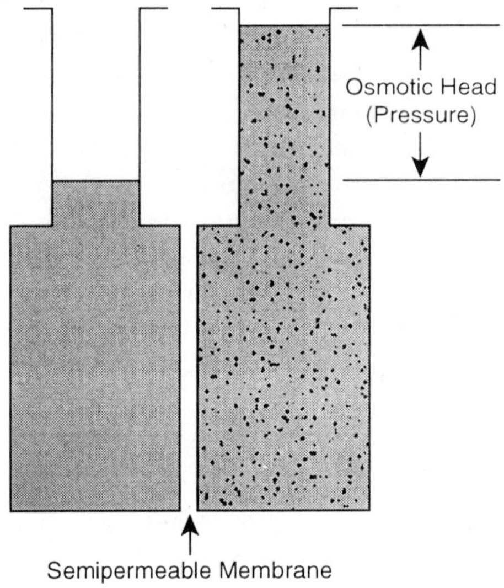

Figure 1-7 Osmotic equilibrium

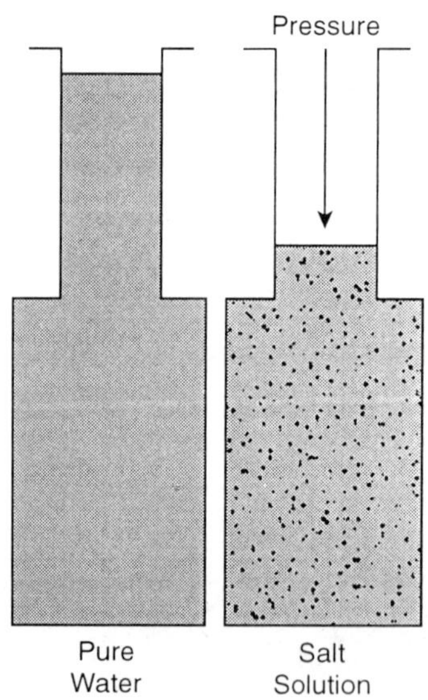

Figure 1-8 Reverse osmosis

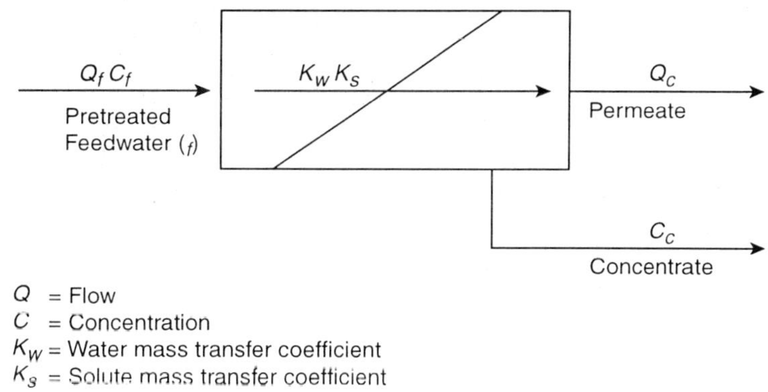

Figure 1-9 Schematic diagram of a reverse osmosis membrane element

certain simplifying assumptions. However, the equations are superior to using percent rejection as a means of predicting permeate water quality.

Figure 1-9 shows a single membrane element, as well as the terminology commonly used to describe flow, concentration, and pressure. The following equations are commonly used for RO and NF membrane processes to determine system efficiency.

Mass balance for water flow:

$$Q_f = Q_p + Q_c \tag{1-1}$$

Where:

Q_f = feedwater flow rate, units of volume per time, e.g., gpd
Q_p = permeate flow rate, units of volume per time, e.g., gpd
Q_c = concentrate flow rate, units of volume per time, e.g., gpd

Mass balance for solute flux:

$$Q_f C_f = Q_p C_p + Q_c C_c \tag{1-2}$$

Where:

C_f = feedwater solute concentration, units of mass per volume, e.g., mg/L
C_p = permeate solute concentration, units of mass per volume, e.g., mg/L
C_c = concentrate solute concentration, units of mass per volume, e.g., mg/L

Product recovery rate:

$$R = \frac{Q_p}{Q_f} \tag{1-3}$$

Where:

R = decimal fraction of product water recovered from feedwater

Water flux:

$$F_w = K_w(\Delta P - \Delta \pi) = \frac{Q_p}{A} \tag{1-4}$$

Where:

F_w = water flux, units of flow rate per unit area, e.g., gpd/ft^2

K_w = water mass transfer coefficient (see Eq 1-6), units of flux per pressure, e.g., gpd/ft^2/psi
ΔP = transmembrane pressure differential, units of pressure, e.g., psi
$\Delta\pi$ = transmembrane osmotic pressure differential, units of pressure, e.g., psi
A = effective membrane area

NOTE: $\Delta P - \Delta\pi$ = the net applied pressure (NAP).

Dissolved solute flux:

$$F_s = K_s(C_m - C_p) = K_s\left[\left(\frac{C_f + C_c}{2}\right) - C_p\right] = \frac{Q_p C_p}{A} \qquad (1\text{-}5)$$

Where:

F_s = solute flux, units of mass per time per area, e.g., lb/ft^2/d
K_s = solute mass transfer coefficient (see Eq 1-7), units of length per time, e.g., ft/d
C_m = concentration at the membrane surface

NOTE: The use of average feed-side solute concentration disregards concentration polarization and film theory effects.

Water mass transfer coefficient:

$$K_w = \frac{Q_p}{A(\Delta P - \Delta\pi)} = \frac{Q_p}{A(NAP)} \qquad (1\text{-}6)$$

Solute mass transfer coefficient:

$$K_s = \frac{Q_p C_p}{A \Delta C} \qquad (1\text{-}7)$$

Where:

ΔC = concentration differential (see Eq 1-8)

Concentration differential:

$$\Delta C = (C_m - C_p) = \left(\frac{C_f + C_c}{2}\right) - C_p \qquad (1\text{-}8)$$

Equation 1-9 describes the effects of each of the five independent variables (K, C_f, K_w, ΔP, and R) on the permeate concentration:

Permeate concentration:

$$C_p = \frac{K_s C_f}{K_w \Delta P + \Delta\pi\left(\frac{2 - 2R}{2R}\right) + K_s} \qquad (1\text{-}9)$$

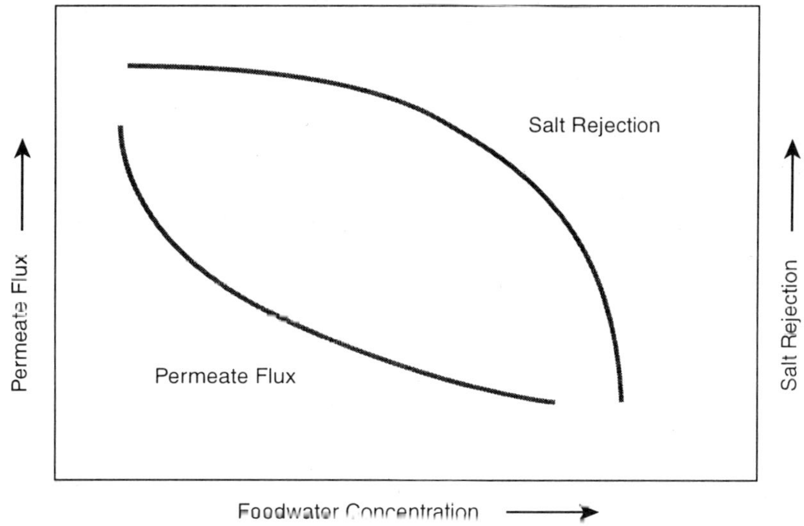

Figure 1-10 Effect of increasing salt concentration on flux and salt rejection

Each of these variables may be changed independently while the other four remain constant. For example, the feedwater may be altered by source selection or pretreatment. The membrane's SR varies inversely with feedwater concentration (Figure 1-10). Thus, C_P will increase if C_f increases and decrease if C_f decreases. The mass transfer coefficients K_w and K_s depend on the membrane. Different membranes have different mass transfer characteristics, a fact that highlights the value of pilot testing membranes before designing a membrane system. Permeate concentration will increase as K_s increases and decrease as K_s decreases. In contrast, C_p will decrease as K_w increases and increase as K_w decreases. Thus, to produce a low-TDS water inexpensively, the best choice is a membrane with a very high K_w and a very low K_s.

Variations in pressure, recovery, and temperature also affect permeate concentration. As pressure increases, the HSD model predicts that permeate concentration will decrease; as pressure decreases, permeate concentration will increase (Figure 1-11). Mass transfer of solutes is diffusion controlled (i.e., independent of pressure) and will progress through the membrane at a constant rate. Pressure can be increased or decreased to affect permeate concentration and recovery. As pressure is increased, more water will be forced through the membrane and the permeate will be diluted. The opposite is true for decreasing pressure.

Recovery variations have the opposite effect on C_P when compared to pressure changes. As recovery increases or decreases, the permeate concentration increases or decreases, respectively (Figure 1-12). If temperature increases and all other parameters are constant, the permeate flux and passage of salt increase (i.e., there is a lower SR), as shown in Figure 1-13. Percent salt passage refers to the percentage of source water TDS or other dissolved solutes of interest that pass through the membranes into the product water. Table 1-3 summarizes the effects of increasing pressure, temperature, recovery, and feed salt concentration on permeate flow and salt passage.

Organics, such as high-molecular-weight DBP precursors, are usually removed by sieving. One method of determining the rejection mechanism is to vary pressure and recovery and observe solute rejection. If the percent solute rejection does not

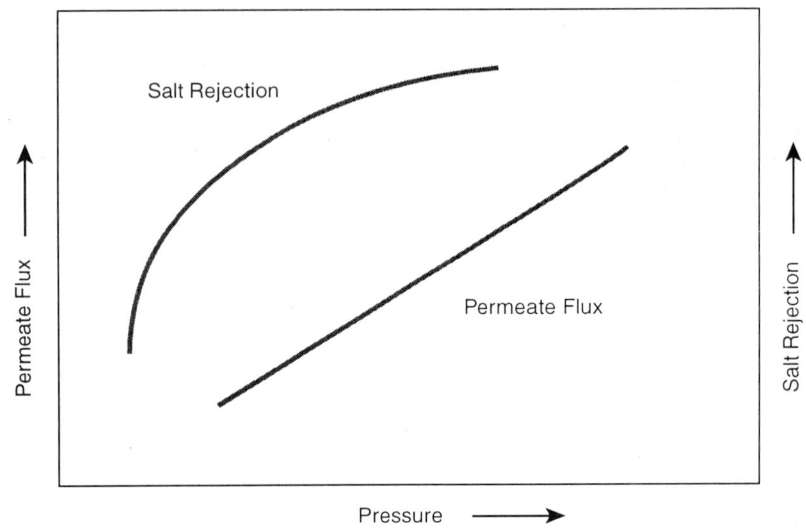

Figure 1-11 Effect of feedwater pressure on flux and salt rejection

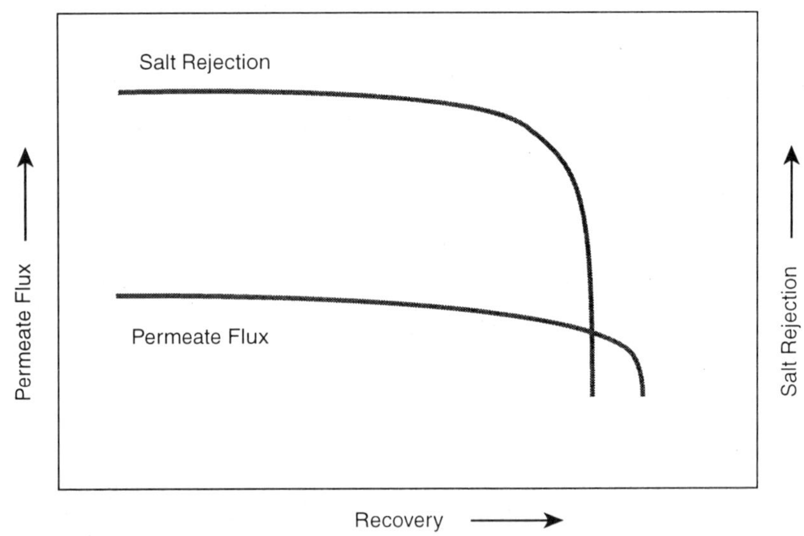

Figure 1-12 Effect of increased recovery on flux and salt rejection

vary with pressure and recovery, sieving is the likely rejection mechanism. If percent rejection varies with pressure and recovery as described in the HSD model, diffusion is the rejection mechanism. Optimizing the process for rejection is easier if the mass transport mechanism is identified.

RATING REVERSE OSMOSIS AND NANOFILTRATION ELEMENTS

All RO and NF devices are rated in terms of initial capacity and SR under a set of standard conditions that are defined by the individual membrane manufacturers. Table 1-4 gives typical standard conditions for a number of RO devices.

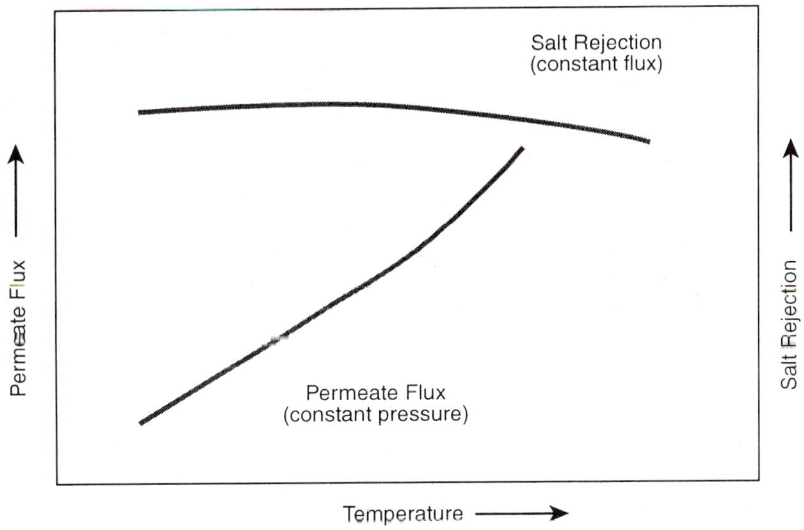

Figure 1-13 Effect of feedwater temperature on flux and salt rejection

Table 1-3 Factors influencing reverse osmosis and nanofiltration performance

Factor	Permeate Flow	Salt Passage
Increasing effective pressure	increases	decreases
Increasing temperature	increases	increases
Increasing recovery	decreases	increases
Increasing source water salt concentration	decreases	increases

Table 1-4 Typical standardized test conditions for various reverse osmosis devices (8-in. diameter modules)

	Brackish				Seawater		
Water Source Configuration	Composite Spiral	Spiral	Hollow Fiber	Hollow Fiber	Composite Spiral	Hollow Fiber	Hollow Fiber
Membrane*	PA	CA	PA	PA	PA	PA	CA
Feedwater composition, *mg/L NaCl*	2,000	2,000	1,500	1,500	32,000	35,000	35,000
Pressure, *psi*	225	420	400	225	800	1,000	780
Temperature, °C	25	25	25	25	25	25	25
Conversion, %	15	10	75	50	8	35	30

*Membrane types: PA, polyamide; CA, cellulose acetate.

The differences in standard conditions make it difficult to compare competitive devices. Valid comparisons can be made only when competitive membranes are evaluated for a common set of operating conditions. Evaluations of RO membranes become feasible only when each device's performance is rated at an equivalent feedwater concentration, temperature, pressure, and recovery. The RO and NF manufacturer's standard conditions should be adjusted to site-specific operating criteria. Because the interrelationships of the equations in this chapter are complicated and time-consuming to calculate, membrane manufacturers use computer software to accomplish the task rapidly. These computer programs define the number and design of RO or NF elements required to deliver a specific quantity of water in a given time under site-specific operating parameters. These programs also predict the average permeate quality obtained from a grouping of RO devices, whether this grouping is the entire plant or a stage in a multistage facility.

In a given system, the number of RO modules needed to produce a specified flow and permeate quality differs for each supplier. The difference in quality is relevant only to the product water specifications of the end user. Appendix C provides information on converting from a manufacturer's standard conditions to site-specific conditions.

REFERENCES

Bailer, J.H., B.W. Lykins Jr., C.A. Fronk, and S.J. Kramer. 1987. Using Reverse Osmosis to Remove Agricultural Chemicals from Groundwater. *Jour. AWWA*, 79(8):55–60.

Bergman, R.A., and J.C. Lozier. 1990. Expanding Applications for Membrane Processes in Water Treatment. In *Proc. of the National Water Supply Improvement Association Biennial Conference.* August 19–23. Lake Buena Vista, Fla.: National Water Supply Improvement Association.

Brunswick, R.J., W. Suratt, and J.E. Burke. 1993. Pilot Testing RO Membranes for Nitrate Removal. In *Proc. of the AWWA Membrane Technology Conference.* Denver, Colo.: American Water Works Association.

Clark, S.W., and M.J. Parrotta. 1991. Membrane Technologies and Drinking Water Regulations. In *Proc. of the AWWA Membrane Technologies in the Water Industry Conference.* Denver, Colo.: American Water Works Association.

Côté, P., M.M. Bourbigot, and K. Agbekodo. 1993. Nanofiltration: An Advanced Process for the Production of High Quality Drinking Water. In *Proc. of the AWWA Membrane Technology Conference.* Denver, Colo.: American Water Works Association.

Filmtec Corporation. 1988. *FT-30 Reverse Osmosis Membrane Specifications.* Technical Bulletin. Midland, Mich.: Dow Chemical Company.

Hofman, J.A.M.H., Th.H.M. Noij, J.C. Kruithof, and J.C. Schippers. 1993. Removal of Pesticides with Nanofiltration. In *Proc. of the AWWA Membrane Technology Conference.* Denver, Colo.: American Water Works Association.

Lozier, J.C., T. McKim, and J. Rose. 1994. Meeting Stringent Surface Water Discharge Requirements for Reclaimed Water With Membrane Processes. Paper presented at the ASCE/NCEE Conference July 11–13 at Boulder, Colo.

Lykins, B.W., et al. 1988. *Reverse Osmosis for Removing Organics From Drinking Water: A Cost and Performance Evaluation.* USEPA Publication PB88–225016. Cincinnati, Ohio: US Environmental Protection Agency.

Mickley, M., R. Hamilton, L. Gallegos, and J. Truesdall. 1994. *Membrane Concentrate Disposal.* Denver, Colo.: American Water Works Association Research Foundation and American Water Works Association.

Taylor, J.S., S.J. Duranceau, W.M. Barrett, and J.F. Goigel. 1989a. *Assessment of Potable Water Membrane Applications and Research Needs.* Denver, Colo.: American Water Works Association Research Foundation and American Water Works Association.

Taylor, J.S., L.A. Mulford, S.J. Duranceau, and W.M. Barrett. 1989b. Cost and Performance of a Membrane Pilot Plant. *Jour. AWWA,* 81(11):52–60.

Toray Industries. 1989. *Toray "Romembra" UTC Membrane Reverse Osmosis Module (SU-Series, SU-710L/720L).* Technical Bulletin Series. China 279, Japan: Toray Industries.

Weber, W.A. 1972. *Physicochemical Processes for Water Quality Control.* New York: John Wiley and Sons.

Chapter 2

Design

This chapter discusses the general design principles of reverse osmosis (RO) and nanofiltration (NF) membrane systems, including the following topics:

- design fundamentals
- water supply
- pretreatment
- process technology
- posttreatment
- concentrate disposal

This chapter is intended to introduce the most important design considerations, but not to an in-depth and exhaustive degree. Reference sources are available for further study in the Additional Sources of Information section at the end of the book.

DESIGN FUNDAMENTALS

Design fundamentals used in selecting and sizing pretreatment, membrane, and posttreatment components of an RO or NF system are discussed in this section, as are source water quality and quantity factors. Refer to Figure 1-3 for a schematic view of the typical membrane system used in the subsequent discussions. Table 2-1 summarizes the typical design factors and related unit processes.

To illustrate these design fundamentals, a hypothetical water system will serve as an example. The design elements for this system are

- brackish groundwater source with water quality as given in Table 2-2
- plant-finished water demand of 1.0 mgd with no blending of source water with membrane product water
- spiral-wound membrane elements

Table 2-1 Membrane system design factors and unit processes

Design Factors	Related Unit Processes
Source water supply	
Water quality	Pumping
Available quantity	Pumping
Pretreatment	
Fouling potential	Filtration
Scaling potential	Acid and antiscalant addition
Membrane process	
Feed pressure	Membrane treatment
Permeate recovery rate	Membrane treatment
Permeate flux	Membrane treatment
Permeate quality	Membrane treatment
Posttreatment	
Stabilization	Aeration–degasification, alkalinity recovery
Disinfection	Disinfection

Table 2-2 Assumed initial water quality for hypothetical system

Parameter	Concentration, *mg/L*
Cations	
Barium	0.04
Calcium	8
Iron	0.5
Magnesium	2
Manganese	0.02
Potassium	8
Sodium	695
Strontium	2
Anions	
Bicarbonate	631
Carbonate	3
Chloride	730
Fluoride	1.1
Orthophosphate	0.7
Sulfate	79
Other	
pH	8.0*
Silicate	24

*Dimensionless value (not milligrams per litre).

Source Water Supply

It is necessary to perform an accurate source water quality analysis of parameters such as those listed in Table 2-3. The water quality analysis allows the designer to correctly select the type and size of the pretreatment and membrane components. The analysis is also used to predict membrane permeate quality. Samples should be split among several certified laboratories with experience in analyzing source water for membrane process treatment in order to rule out possible errors. Critical parameters, such as barium, strontium, sulfate, and fluoride, must be analyzed accurately at appropriate detection levels to evaluate scaling potential, particularly for RO applications. For many NF applications, fouling potential—as generally indicated by dissolved organic carbon (DOC),

Table 2-3 Suggested parameters for water quality analysis*

Cations	Anions	Other
Aluminum	Bicarbonate	Alkalinity
Ammonium	Carbonate	Boron
Barium	Chloride	Color
Calcium	Fluoride	Conductivity
Chromium	Nitrate	Dissolved oxygen
		Heterotrophic plate count
Copper	Sulfate	Hydrogen sulfide
Iron (ferrous and ferric)	Phosphate (total and ortho)	Ionic strength and balance (calculated from analyses)
Lead		Organic carbon (total and dissolved)
Magnesium		pH
Manganese		Silica (total and colloidal)
Potassium		Silicon
Sodium		Silt density index
Strontium		Sulfide-generating bacteria
Zinc		Temperature
		Total coliform
		Total dissolved solids
		Total hardness
		Turbidity
		Yeast and mold

*Not all of these analyses are required for all membrane design calculations. Membrane manufacturers should be consulted for a recommended list of water quality parameters to be tested.

turbidity, and/or biological activity indicators such as heterotrophic plate count—is a critical analytical need. The reader is cautioned that fouling rates should be determined based on pilot plant tests for the most accurate indication of fouling potential.

Of primary interest are the following concentrations:

- major cations, such as sodium for brackish saline waters (an RO application) and calcium and magnesium for membrane softening (an NF application)
- major anions, such as chloride and sulfates for brackish and saline waters
- major and minor cations and anions, as well as silica, that can form relatively insoluble salts (limiting salts) that can limit permeate recovery
- potential foulants, including metal hydroxides, silicates, iron and manganese salts, bacteria, algae, and colloidal material
- hydrogen sulfide

The required feedwater flow rate will exceed the permeate production because of the limitations in permeate recovery. Assuming typical permeate recovery rates ranging from 50 to 90 percent, the ratio of required source water to permeate could range from 1.11 to 2.00 as shown in Table 2-4. Figure 2-1 shows the required feedwater flow rates for facilities with capacities of 1.0 mgd or greater at various permeate recovery rates. This additional flow should be considered in the design of wells, pumps, and delivery piping. Also, if two or more water sources of different quality are available, an analysis of membrane process costs versus source water supply costs should be performed. It is also very important to assess the chemical stability of the source water.

Table 2-4 Feedwater flow rates based on permeate recovery rates

Permeate Flow Rate, *mgd*	Product Recovery Rate, %	Feedwater Flow Rate, *mgd*	Concentrate Water Flow Rate, *mgd*
1.00	50	2.00	1.00
1.00	60	1.67	0.67
1.00	70	1.43	0.43
1.00	75	1.33	0.33
1.00	80	1.25	0.25
1.00	85	1.18	0.18
1.00	90	1.11	0.11

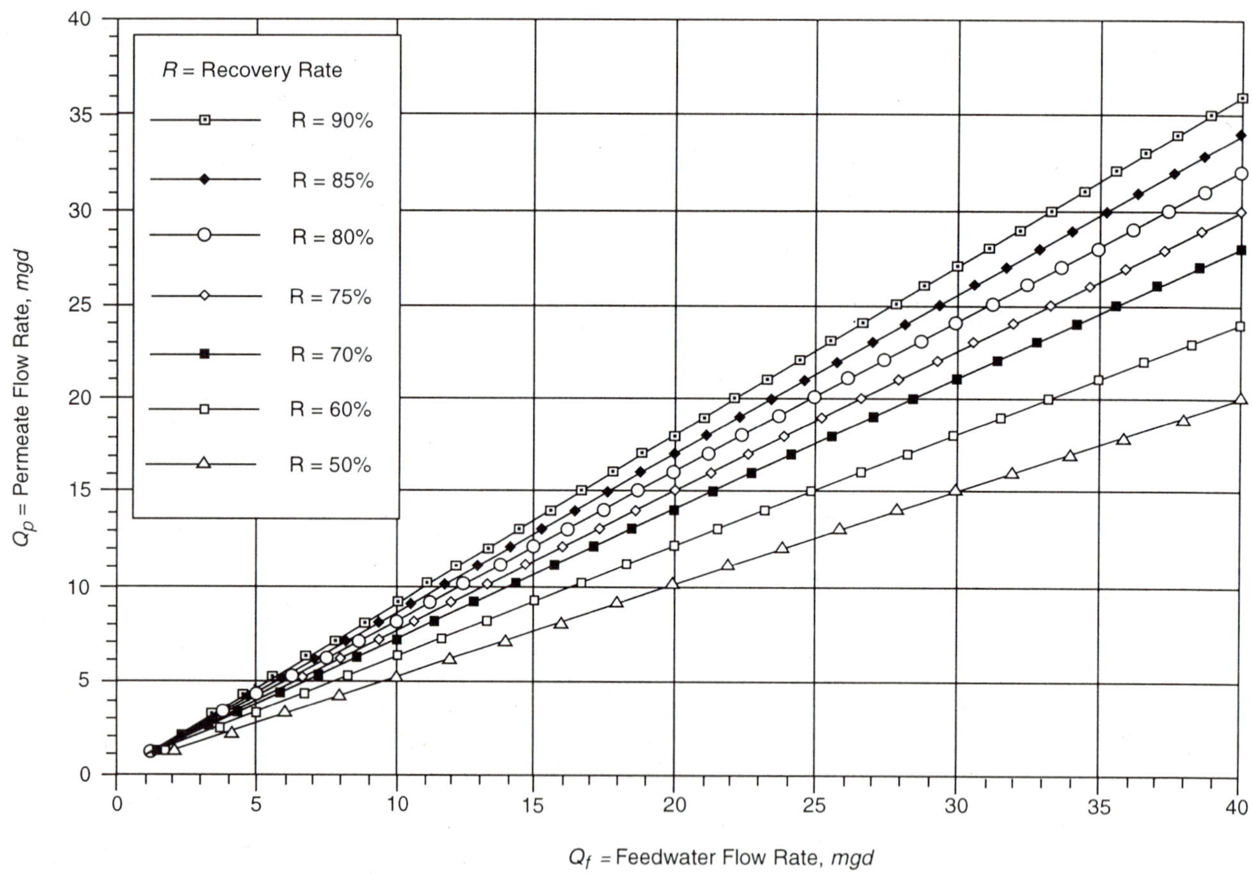

Figure 2-1 Feedwater flow rates for various permeate recovery rates

Pretreatment

The two major processes generally used for pretreatment are scale control and prefiltration. In some instances, additional treatments are required to remove algae; certain metals, such as iron and manganese; or other particulate matter. The need for such treatment is dictated by the source water analysis, the permeate standards, and the need to protect the membranes. Only the most commonly encountered contaminants are discussed in this manual.

Scale control. As product water that is relatively low in solutes passes through the membrane, the remaining feedwater becomes increasingly concentrated in dissolved solids. At certain degrees of concentration, the solubility of various salts can be exceeded, which causes salts to precipitate onto the membrane surface. This process is called *scaling*. Scaling can severely reduce the flow or flux of permeate and can irreversibly damage the membrane. Calcium sulfate is particularly notorious for its ability to damage membranes.

For a large proportion of mineralized source waters, the concentration of salts in the concentrate stream (and conversely the degree of permeate recovery) is limited to the point at which the membrane process becomes uneconomical without pretreatment. This situation is particularly true for water with high concentrations of calcium, barium, strontium, sulfates, or carbonates. With silica precipitation, however, a certain degree of oversaturation is generally tolerated because the chemical reaction is relatively slow and feedwater leaves the system before the precipitate forms. However, membrane manufacturers or system suppliers should always be consulted before silica oversaturation is allowed for a membrane system. Lime softening at a pH greater than 11.0 can be used to reduce silica. Inhibitors to prevent silica oversaturation also show promise in addressing this problem. Table 2-5 provides a list of common limiting salts or compounds.

Limiting salt calculations are normally performed with computer software. Appendix C illustrates the general steps used to determine the limiting salt and calculate allowable permeate recoveries. The same appendix also gives examples using the data from Table 2-2. The results of those calculations show that, for the hypothetical feedwater, the solubility of calcium carbonate is exceeded and calcium carbonate will precipitate. The results also show that a recovery rate of 84 percent is possible before the solubility limit of strontium sulfate is exceeded. The examples in appendix C are illustrative and do not take the ionic strength of the solution into account. The ionic strength of the solution will make the salt more soluble and, in effect, increase the allowable permeate recovery rate. Design software that membrane manufacturers use takes into account the impacts of ionic strength in determining allowable recovery rates.

To increase the permeate recovery rate, feedwater may be altered chemically to reduce the concentration of cations or anions of the limiting salt. One method to prevent the formation of calcium carbonate scales is to add an acid to convert carbonate ions (CO_3^{2-}) to bicarbonate ions (HCO_3^-). Bicarbonate ions can also be converted to carbonic acid (H_2CO_3) and subsequently carbon dioxide during acidification. Appendix C gives an example of adding sulfuric acid to control calcium carbonate scaling.

Antiscalants are often added to the feedwater stream to allow the membrane system to operate at permeate recovery rates in excess of limiting salt constraints. Antiscalants inhibit the rate of formation of crystals by interacting with the

Table 2-5 Common limiting salts

Salt	Solubility product of salt, pK_{sp} at 25°C
Calcium carbonate ($CaCO_3$)	8.3
Calcium sulfate ($CaSO_4$)	4.7
Barium sulfate ($BaSO_4$)	9.7
Strontium sulfate ($SrSO_4$)	6.2
Silica (SiO_2)	2.7

compounds as they start to form crystals. As the crystals begin to form on the membrane surface, the feedwater velocity is sufficient to break up the salt crystals and flush the particles from the membrane surface. Antiscalant dosage calculations are too complex to be presented in this manual. Chemical dosage and efficacy information for specific membranes is available from antiscalant manufacturers, and membrane manufacturers provide information on the compatibility of antiscalants with their membranes.

Besides acids or antiscalants, other methods of controlling scaling or colloidal fouling include

- ion exchange softening to reduce multivalent cations (Ca^{2+}, Mg^{2+})
- lime softening to reduce multivalent cations and silica
- aeration and filtration to reduce iron and manganese
- "greensand" or oxidation filtration to reduce iron and manganese
- coagulation and filtration to reduce colloidal and particulate matter

For the first four methods, standard texts on potable water treatment can be consulted. Colloidal and particulate fouling is discussed in this manual.

Prefiltration. Essentially all membrane systems use micron-sized prefilters to protect the downstream membrane from fouling and from mechanical damage caused by particulate matter. Such particulate matter can include sand and rust particles and other corrosion by-products. Prefiltration is usually accomplished by using cartridge filters in pressure vessels.

Cartridge filter pore size is usually 5 microns, with a typical range of 1 to 25 microns. Cartridge filters are usually selected for a maximum flow of 5 gpm per 10 in. of cartridge filter length, and they are commonly available in lengths ranging from 10 to 40 in. The typical length is 40 in. for membrane systems.

To size the total number of cartridge filters or filtration area required, the designer must also account for the permeate recovery rate and excess feedwater. Thus, a 1.0-mgd permeate capacity facility operating at a 75 percent permeate capacity rate will require the prefiltration of 1.33 mgd of source water. Based on a maximum flow of 5.0 gpm per 10 in. of cartridge filter, a total of approximately 46, 40-in.-long cartridge filters would be required to filter 1.33 mgd of source water. This flow rate per cartridge filter length should be considered a maximum. To be safe, the actual design size should range from 3 to 4 gpm per 10-in. length, depending on the application. Additional filtration capacity is normally provided to allow cartridges to be replaced without the need to interrupt production. Pressure vessels, usually constructed of 316 stainless steel, can hold up to 704 cartridge elements for a flow of 3,520 gpm per vessel.

The clean-filter pressure drop is usually specified as less than 3 psi. It is necessary to replace a filter cartridge when the transfilter pressure drop reaches 15 psi. The operational time before replacement depends on source water quality and the degree of pretreatment. Typically, a cartridge filter changeout is recommended once every two to four months.

Membrane Array Design

The term *membrane array* refers to a series of membrane elements, usually arranged in stages with a decreasing number of membrane elements in each succeeding stage. An example is a three-stage system with the number of elements in a ratio of 4:2:1 going from stage 1 to stage 2 to stage 3. Figure 2-2 illustrates a three-stage array. (The purpose of the stages will be discussed later in this chapter.) When the feedwater and finished water characteristics have been measured or otherwise

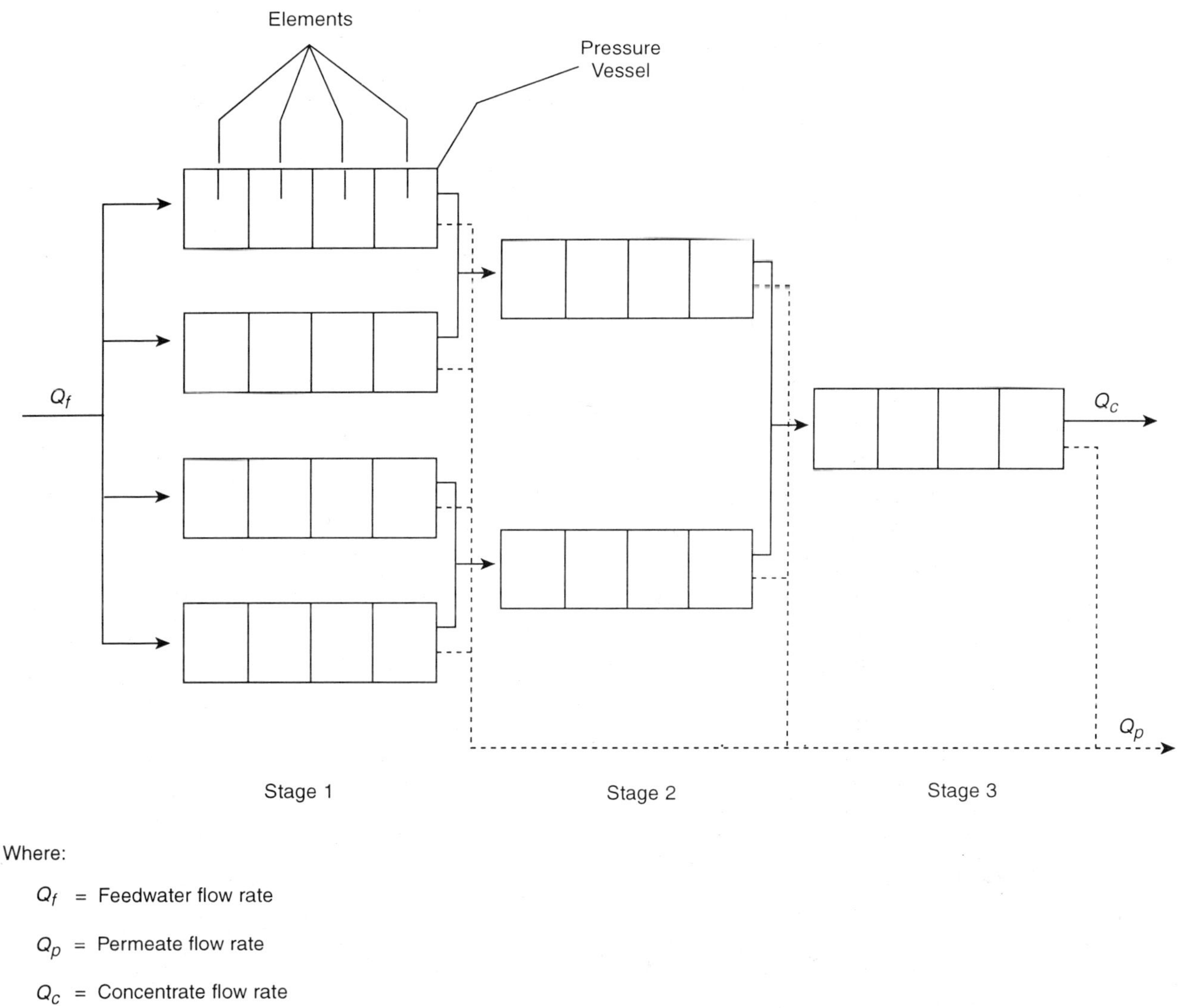

Figure 2-2 Schematic diagram of a reverse osmosis membrane device

determined, the actual membrane system or array can be designed. The reader should consult appendix C for more information. The process involves the following steps:

1. Determine the appropriate membrane process (NF, RO, etc.).
2. Set maximum flux rates and recovery rates (total and per membrane element or pressure vessel).
3. Determine the net applied pressure limits.
4. Determine the solute rejection requirements.
5. Select the membrane type.
6. Reassess and set design flux, pressure, and recovery rates for optimum treatment and cost.
7. Determine the preferred membrane train number and size.

8. Establish the amount of necessary membrane area (spiral-wound membranes only).
9. Select the membrane element diameter and length.
10. Select the number of elements per pressure vessel (spiral-wound membranes only).
11. Determine the number of stages consistent with selected per element recovery rate.
12. Calculate the array size consistent with membrane train as well as number of elements and pressure vessels.
13. Calculate the predicted permeate solute concentration and assess against target value.
14. Evaluate the potential ability to blend feedwater with source water and repeat steps 6 through 13, as necessary.

The following discussion provides an overview of these membrane array designs.

Membrane behavior. A number of variables determine the overall water production and permeate quality. The important design equations can be found in the theory section of chapter 1. These equations are used to calculate permeate recovery rate, water mass balance, solute mass balance, water flux, solute flux, and permeate solute concentration.

Recovery. The permeate recovery rate R is the ratio of the permeate flow rate to the feedwater flow rate. Generally the permeate recovery rate is set at the highest possible rate consistent with the desired water quality concerns, feedwater scaling limits, and lowest capital and operating costs. Incorporating solute concentration with each flow stream allows a mass balance to be established for the membrane system.

Water flux. The water flux is controlled by the water permeation coefficient (K_w, also called the water mass transfer coefficient), the water temperature, and the net driving pressure. The water permeation coefficient is inherent to a particular membrane and is based on the materials and conditions of construction. Net applied pressure is the difference between the feed-side pressure and permeate-side pressure. Net driving pressure is equal to the net applied pressure minus the average or mean osmotic pressure. The difference between the feed-side and permeate-side applied pressure selected by the designer can be varied over a relatively wide range within the mechanical constraints of the membranes. The water flux typically increases by about 3 percent per degree Farenheit temperature increase.

Solute flux. Solute flux depends on the concentration difference across the membrane and the solute permeation coefficient (K_s, also called the *solute mass transfer coefficient*). The solute permeation coefficient is inherent to the membrane material and condition of construction. The concentration difference can be selected by altering the permeate recovery rate or by using a different source water.

Permeate solute concentration. Equation 1-9 (p. 16) is used to predict permeate solute concentration. The equation incorporates the basic equations for solute and water flux and is based on five independent variables: feedwater solute concentration, solute permeation coefficient, water permeation coefficient, permeate recovery rate, and net applied pressure. Note that net driving pressure is also commonly referred to as *transmembrane pressure differential* and is occasionally referenced as such in this manual. The feedwater solute concentration is inherent to the selected supply source. The solute and water permeation coefficients are inherent

Table 2-6 Assumptions for the permeate solute concentration example

Parameter	Value
Permeate design flow	1.0 mgd
Membrane element area	80 ft^2
Feedwater TDS concentration	1,000 mg/L
Water flux values	8 gpd/ft^2, 12 gpd/ft^2, and 16 gpd/ft^2
Net applied pressures	100 psi, 250 psi, and 500 psi
Water permeation coefficients	0.005 day^{-1}, 0.05 day^{-1}, and 0.5 day^{-1}
Solute permeation coefficients	0.1 ft/d, 1.0 ft/d, and 10.0 ft/d
Permeate recovery rates	50%, 75%, and 90%

Table 2-7 Impact of varying permeate recovery rate*

Permeate Recovery Rate, %	Feedwater TDS Concentration, *mg/L*	Concentrate Water TDS Concentration, C_c, *mg/L*	Average Feedwater Side, *mg/L*	Permeate TDS Concentration, C_p, *mg/L*	Required Feedwater Flow Rate, *mgd*
50	1,000	1,951	1,475	49	2.00
75	1,000	3,762	2,381	79	1.33
90	1,000	8,566	4,783	159	1.11

*These values assume C_p = 1.00 mgd, K_w = 0.05 day^{-1}, K_s = 1.0 ft/d, ΔP = 250 psi.

to a particular membrane. Both the permeate recovery rate and net applied pressure can be set by the designer or operator.

To illustrate the impacts of changes in the variables on permeate solute concentration, the example presented here uses the design parameters in Table 2-6.

Impact of varying permeate recovery rate. Table 2-7 shows how varying the permeate recovery rate significantly changes the permeate (C_p) and concentrate (C_c) solute concentration. The table also shows the significant decrease in feedwater flow rate with increasing permeate recovery rates.

Impact of varying feedwater pressure. Table 2-8 shows the variation in C_p with a change in net applied pressure. Comparing Tables 2-7 and 2-8 demonstrates that a 50 percent increase in permeate recovery rate (from 50 percent to 75 percent) results in a 67 percent increase in C_p, while a 50 percent decrease in net applied pressure (from 500 psi to 250 psi) results in a 93 percent increase in C_p. Thus, net applied pressure has a greater impact per unit change on C_p than permeate recovery rate.

Figure 2-3 illustrates how permeate solute concentration can vary with net applied pressure and permeate recovery rate. A sharp rise in the solute concentration occurs with low applied pressure and high permeate recovery rate.

Impact of varying membrane characteristics. Table 2-9 shows the relative impact of varying the membrane's water and solute permeation characteristics (K_w and K_s, respectively). Extreme differences in permeate solute concentrations can be realized by using different membrane materials and construction methods.

Water flux rate. Permeate flow can be varied by changing the water permeation coefficient and net applied pressure. Based on the assumed design parameters in Table 2-6, the relative impacts on water flux (F_w) achieved by manipulating these variables can be demonstrated.

Table 2-8 Impact of varying net applied pressure*

Net Applied Pressure, *psi*	Permeate TDS Concentration, C_p, *mg/L*
100	177
250	79
500	41

*These values assume C_f = 1,000 mg/L TDS, K_w = 0.05 day $^{-1}$, K_s = 1.00 ft/d, R = 75 percent.

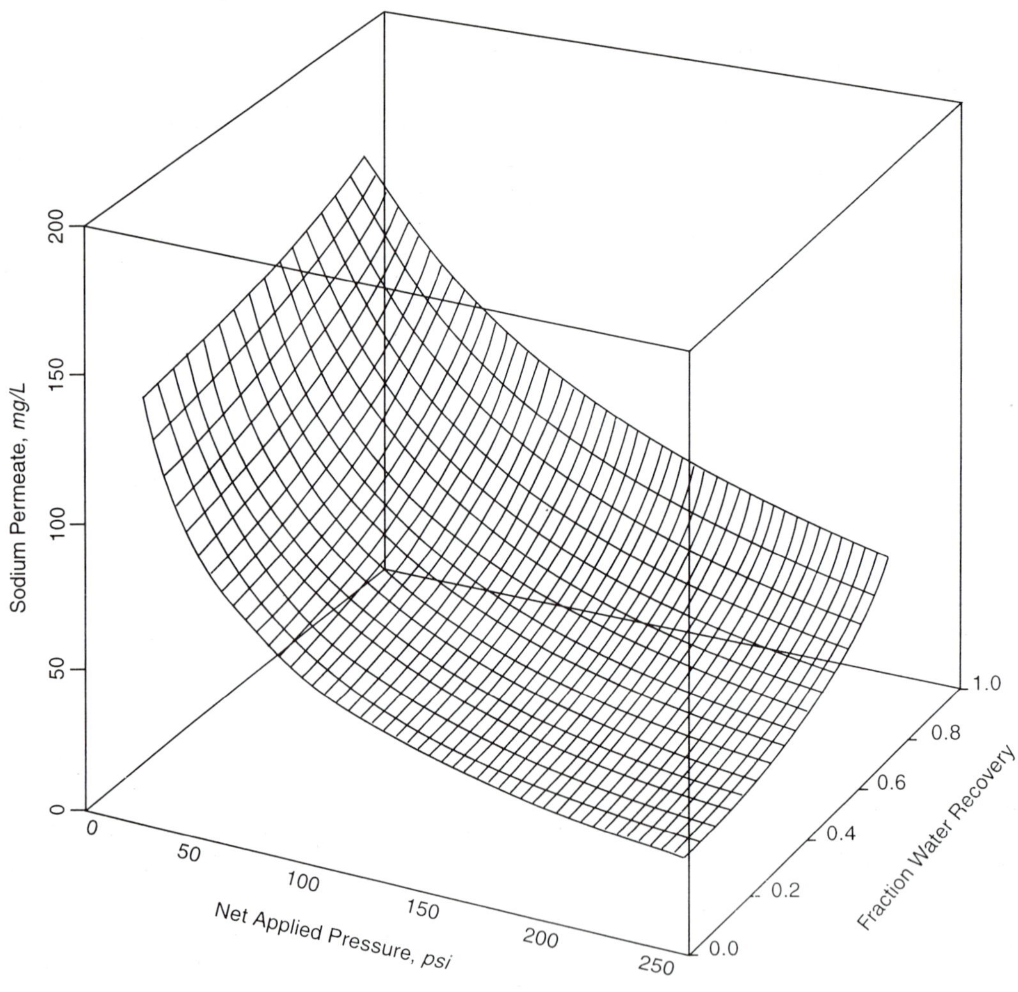

Figure 2-3 Three-dimensional plot of permeate sodium concentration as a function of net applied pressure and permeate recovery rate

Impact of varying net applied pressure. Table 2-10 demonstrates the relative impact of varying net applied pressure on water flux. The response of flux to net applied pressure is directly proportional in a linear fashion.

Impact of varying water permeation coefficient. Table 2-11 demonstrates the variation in water flux achievable by selecting membranes with differing water permeation coefficients. Again, the impact is directly proportional in a linear fashion.

Table 2-9 Impact of varying membrane characteristics on C_p*

K_s Value, ft/day	C_p, mg/L		
	K_w = 0.005 day^{-1}	K_w = 0.05 day^{-1}	K_w = 0.5 day^{-1}
0.1	79	9	0.9
1.0	463	79	9
10.0	896	463	79

*These values assume R = 75 percent, ΔP = 250 psi, C_F = 1,000 mg/L TDS.

Table 2-10 Impact of varying net applied pressure on water flux*

Net Applied Pressure, *psi*	Water Flux (F_w), *ft/d*
100	11.6
250	29.0
500	58.1

*These values assume K_w = 0.05 day^{-1}

Table 2-11 Impact of varying water permeation coefficient on water flux*

Water Permeation Coefficient (K_w), *day^{-1}*	Water Flux (F_w), *ft/d*
0.005	2.90
0.05	29.0
0.5	290

*These values assume ΔP = 250 psi.

Membrane array configuration. Typically, an assumed maximum flux is set during the design process to prevent undue fouling rates. This assumed maximum water flux ranges up to 16 gpd/ft^2 for spiral-wound membrane elements for fresh and brackish sources. Note that water flux is often referred to in feet per day (ft/d). Units of gpd/ft^2 may be converted to units of ft/d by dividing by 7.48. Hollow fiber membrane fluxes are less than 1 gpd/ft^2. In general, membrane train sizes are designed in increments of 0.5 mgd or higher, even though target flows can be matched to approximately 5,000 to 10,000 gpd. Train sizes are generally limited to 2.5 mgd. Membrane arrays are generally one to three stages with multiple membrane elements connected in series with each stage (refer to Figure 2-2). Generally, the permeate recovery rates for a one-, two-, or three-stage array with six 40-in.-long membrane elements per pressure vessel assembly are as follows:

- one stage: ≤ 50 percent
- two stages: > 50 percent and ≤ 75 percent
- three stages: > 75 percent

For spiral-wound membranes, depending on feedwater quality, two-stage arrays can achieve recoveries of more than 75 percent by using seven elements per pressure vessel (i.e., more membranes) or by recycling a portion of the concentrate to a point before the first stage. The number of membrane elements (and associated vessels) in a second or third stage is usually one-half the number of elements in the preceding stage. This decrease in elements is necessary because the preceding stage removes approximately 50 percent of the feedwater entering that stage as permeate. Appendix C provides an example of the process used to arrive at a selected membrane array and train size.

Membrane System Control

Membrane facilities are generally controlled by measuring permeate and concentrate flow rates, then adjusting the applied membrane pressure to maintain a constant process recovery. As membrane systems are operated, membrane fouling requires that a higher pressure gradually be applied to produce the target permeate flow rate. For this reason, membrane feed pumps are designed with additional head capacity over the "clean membrane" condition. A valve on the concentrate water line controls the permeate recovery rate by controlling feed pressure to maintain the set-point permeate recovery of the process. This valve is particularly important during initial operation of the membranes, when the "clean" condition of the membrane would tend to overproduce permeate and exceed the desirable water flux. As membranes age, productivity usually declines, and the concentrate water line control valve can be used to maintain overall product water flow needs.

Most membrane systems designed today use variable-speed motors for the membrane pressure pump to reduce the throttling (and wasting) of feed pressure. Tables 2-12 and 2-13 provide a listing of typical alarms and instruments used in membrane plants. Because the operating data gathered by the instrumentation must be normalized before operating conditions can be evaluated, programmable logic controller based technology is typically used in membrane plants.

Posttreatment

Permeate leaving the membrane system is typically low-pH water with excessive dissolved gases and high corrosion potential. This condition results from the acid used to minimize carbonate and from the significant reduction of calcium, magnesium, and bicarbonate ions. Acid converts bicarbonates to carbon dioxide, which easily passes through the membranes into the permeate, while the remaining bicarbonate ions are readily rejected, especially for the RO process. If it is not removed, carbon dioxide will redissolve to form carbonic acid. In addition, any dissolved gases in the feedwater, such as hydrogen sulfide, also pass through the membrane into the permeate. Thus, the corrosivity of the water is due to a combination of low pH, excessive carbon dioxide, and a lack of buffering capacity (bicarbonates).

Although membranes are highly effective in removing protozoan, bacterial, and viral populations, minute imperfections in the membranes, leaks through the glue lines in the membrane elements, leaks in brine seals, or leaks in hollow fiber element O-rings or permeate tube interconnector O-rings can allow such biological contamination to pass through. Before potable use of the permeate, the posttreatment processes discussed in the following text may be necessary:

Aeration and degasification. A packed tower process using forced aeration strips carbon dioxide and hydrogen sulfide (if present) from the permeate. This process also adds dissolved oxygen to the water, increases the pH, improves taste,

Table 2-12 Typical list of alarms for a brackish water reverse osmosis plant

Stage	Parameter or Device	Condition	Shutdown	Advisory
Pretreatment	Feedwater turbidity	High	X	
	Feedwater pH	High, low, deviation	X	
	Cartridge filter ΔP	High		X
	Acid day tank level	Low		X
	Acid pump	Failure	X	
	Scale inhibitor tank 1 level	Low		X
	Scale inhibitor tank 2 level	Low		X
	Scale inhibitor flow	Low		X
	Scale inhibitor pump	Failure	X	
RO train	High-pressure pump suction pressure	Low		X
	RO bank 1 ΔP	High	X	
	RO bank 2 ΔP	High	X	
	Permeate water flow	Low		X
	Permeate conductivity	High	X	
	Concentrate flow	Low	X	
	Recovery	High	X	
	Inlet valve	Close	X	
	High-pressure pump	Failure	X	
	Permeate dump valve	Open		X
	Stand-by pump	Select		X
	Pump dump valve	Open		X
	Test mode	Select		X
	Manual mode	Select		X
	Rinse mode	Select		X
Posttreatment	Permeate pH	High	X* (NaOH pumps)	
	Permeate pH	Low		X
	Clearwell level	High	X	
	Clearwell level	Low		X
	Sample pump	Failure		X
	Air blower	Failure	X	
	Transfer pump	Failure		X
	Caustic soda day tank level	Low		X
	Caustic soda pump	Failure		X
	Storage tank	High		X
	Storage tank	Low		X
Miscellaneous	Computer	Fault	X	
	Stand-by pump suction pressure	Low	X	
	Stand-by dump valve	Open		X

*Or reduce output.

and minimizes odors. Where hydrogen sulfide is absent and the carbon dioxide concentration is low, this step can be eliminated.

Disinfection. Chlorine is generally used for disinfection because of its recognized efficiency as a disinfectant and because there is a reduced level of disinfection by-product precursors in the permeate. However, other final disinfectants could be used depending on a utility's specific practices.

Stabilization. Using sodium hydroxide to raise the pH and convert carbonic acid to bicarbonate and adding corrosion inhibitors will reduce the corrosivity of the permeate. In certain instances, a portion of disinfected and/or filtered source water can be blended with permeate to increase the alkalinity of finished water.

Alkalinity recovery. Alkalinity can be recovered from the permeate by the addition of caustic soda before aeration. This addition raises the pH of the permeate, converting carbonic acid to bicarbonate, which is not lost in the subsequent aeration. This process is economical for all permeate that has relatively low hydrogen sulfide

Table 2-13 Typical list of panel-mounted instruments for a brackish water reverse osmosis plant

Stage	Parameter	Instrument(s)
Pretreatment	Feedwater temperature	Indicator, recorder
	Feedwater conductivity	Indicator, recorder
	Feedwater turbidity	Indicator, recorder
	Feedwater pH (adjusted)	Indicator, recorder, controller
	Scale inhibitor flow	Controller
RO train	Inlet pressure	Indicator, recorder
	Permeate flow	Indicator, recorder, totalizer
	Concentrate flow	Indicator, recorder, controller, totalizer
	Permeate conductivity	Indicator, recorder
	Concentrate conductivity	Indicator
	Membrane pressure pump (motor)	Hour meter, event counter, current indicator
Posttreatment	Permeate conductivity	Indicator, recorder
	Permeate pH	Indicator, recorder, controller
	Concentrate pH	Indicator, recorder

levels. For high-sulfide waters, removal of hydrogen sulfide by aeration is impeded at higher pH values (>7.0).

SOURCE WATER SUPPLY

One of the most important components of any membrane treatment system is a reliable source water supply (Missimer 1994). Even the most sophisticated, well-designed membrane treatment facility has limitations in its ability to treat highly contaminated or varying quality source water. This section discusses the development of source water supply and maintenance of water quality.

Surface Water

Surface water use presents numerous technical problems for pretreatment systems for membrane treatment processes. Membranes are very sensitive to the concentration of particulate material in the water. A surface water source may require construction of expensive pretreatment facilities to prevent particulate material from plugging the prefilters. Beach or sand wells are sometimes used to avoid direct intakes of surface water.

Fresh water. Intakes for most fresh surface water treatment facilities, regardless of the treatment process, are submerged pipes into rivers, lakes, or reservoirs. The pipes are normally placed at a mid-water depth to avoid the entrance of either floating debris or bottom sediments. A fixed-position intake pipe is commonly found where sufficient data are available to determine that the pipe will always be submerged regardless of climatic conditions. However, many utilities prefer multiple-depth openings to optimize source water quality.

As the number of membrane treatment applications for the removal of organics from essentially fresh water increases, direct surface water intakes are more likely to be used. The elevation of the intake pipe should be at least a few feet below the water surface to avoid inflow of floating debris and higher concentrations of aquatic plants, which tend to stay in the photic zone (i.e., those areas of water that are penetrated by light).

The intake pipe must also be placed as far from the bottom as possible to avoid collection of sediments. One practical design for a surface water intake employs a floating platform to maintain the intake in the middle of the water column, with the pumping facility located at a secure position on land, above an established flood level.

The end of the intake should be screened to avoid the influx of large debris. The diameter of the intake pipe should be sized to allow a proper volume of flow, depending on the need to maintain a low inflow velocity to prevent sediment entrainment and the scouring of vegetative growth from the pipe.

Seawater or brackish water. Intakes from tidal surface water bodies feeding RO water treatment facilities share many common problems with freshwater systems. However, a number of other issues, such as the susceptibility of a facility to corrosion, storm damage, and marine organism growth, must be considered.

A number of existing seawater membrane treatment facilities use surface intake systems. Many of these systems are located in the Middle East or on islands, both natural and artificial (i.e., oil platforms). Most large-scale membrane facilities using direct surface water intakes are located adjacent to large tidal water bodies, such as the Atlantic Ocean or the Arabian (Persian) Gulf. Some critical problems in the design of intakes in these areas are how far offshore and how deep the intake should be placed and what type of terminal should be used.

Oceanographic investigations for intakes are commonly required to assess storm impact potential, potential growth rates of attaching benthic marine organisms (such as corals and sponges), and near-shore water quality fluctuations. The terminal must be below a water depth affected by damaging orbital storm wave motion and yet far enough offshore to avoid the near-shore sediment transport area where storms can cause excavation and suspension of large quantities of sediment. Water quality changes must also be considered in the establishment of the intake point. In the Arabian Gulf, water quality and temperature vary significantly. In some instances, the final intake point may lie several miles offshore. Alternative designs should be considered before a surface intake is installed into any tidal water body.

Intake design. Conventional surface water intakes are susceptible to direct physical damage caused by storms, corrosion, freezing and thawing in temperate climates, and human activities. Unavoidable variations in water quality pose serious problems that may be overcome by designing for worst-case conditions. These variations may be caused by storms, biological activity, temperature changes, seasonal water quality changes due to evaporation during hot, dry periods, and contamination by human-made pollutants. Table 2-14 summarizes surface intake design elements and related considerations.

The considerations listed in Table 2-14 commonly require special steps to be taken in the design or operation of conventional intake systems. Many intake systems have been designed to reduce the concentrations of organic compounds and suspended sediments in source water. Such systems include Ranney collectors, canals with a settling reservoir, groundwater recharge systems with a surface water intake, beach wells or galleries, and seabed filtration systems (rapid sand filters constructed in place). The most effective intake system modification for a given location is highly site-specific.

Groundwater

Under steady-state conditions, groundwater normally provides a chemically stable source of water over a long time. However, in areas adjacent to a coast where pumping takes place, groundwater quality can vary. Overpumping can cause saltwater intrusion, which in turn may cause a membrane treatment plant to fail unless it is designed to accommodate the increased concentration of dissolved solids.

The quality of the surface water supply source can be monitored for many years to assess needed changes in a treatment facility's design. In the development of a

Table 2-14 Surface water intake design elements and considerations

Design Element	Related Considerations
Elevation of terminal	Inflow during low water or drought conditions Inflow during floods or high water Protection from damage, such as erosion or ice damage Minimization of sediment inflow Minimization of floating debris and plant debris inflow
Intake screen	Clogging and maintenance Corrosion
Pipe material	Corrosion Maintenance access
Intake pump	Corrosion Backup Power failure Damage potential (based on location)
Intake location (offshore)	Water depth Water quality variation Storm damage Damage from ship collision

groundwater source, the aquifer must be carefully tested and modeled to determine the anticipated short- and long-term changes in water quality. Under normal pumping conditions, a groundwater system (or aquifer) generally produces a constant quality of water for many years. However, groundwater quality can change depending on the nature of the aquifer used or the presence of overpumping. The potential for significant water quality changes should be considered in the design of the membrane treatment system.

Wellfield design and performance evaluation. Detailed hydrogeologic information is needed before a wellfield can be designed and constructed. The level of sophistication of the test program and subsequent modeling depends on the size of the system and the local hydrogeology. The following questions must be answered before a wellfield is designed:

- Is a groundwater source available?
- How much water will the groundwater source produce (safe yield)?
- Will the water quality remain stable during pumping?
- How many production wells must be constructed to safely obtain the desired yield?
- What are the optimal well locations and pumping rates?
- If water quality changes, what will be the changes in concentration of key chemical parameters?

If the wellfield design is complex and many hydrogeologic data are obtained, a computer model can evaluate both the pumping-induced drawdown of the potentiometric surface and the water quality changes. If the wellfield is in a location not subject to water quality changes and the water essentially serves as source water for a freshwater organics removal application, then a standard hydraulic analysis model should be employed based on a code such as the McDonald and Harbaugh (1988) model.

For wellfields in saline water aquifers or aquifers subject to saltwater intrusion where hydrogeologic data are limited, the two-dimensional solute transport models by Konikow and Bredehoeft (1978) or Sanford and Konikow (1985) can be used. In complex saline water aquifers subject to both horizontal and vertical water quality changes where detailed hydrogeologic data have been collected, the three-dimensional solute transport model SWIFT (Reeves et al. 1986), or an equivalent, is most appropriate. Of course, the modeling output will be only as good as the field data collected and placed into the model.

All of the commercially available solute transport models predict concentrations of substances that do not chemically react in the aquifer or the confining beds. Total dissolved solids (TDS) or dissolved chloride concentrations are commonly modeled and essentially do not chemically react within the production aquifer to change concentrations.

Because membrane treatment facilities are sensitive to certain reactive substances (such as dissolved iron, strontium, silica, or sulfate), special decay equations can be coupled to the standard models to predict concentrations of these substances with respect to time. At the present time, the accuracy of predicting future concentrations of reactive ions is limited. Therefore, if there are potential problems with substances such as iron or sulfate, then the treatment process design should take these problems into account.

A properly designed hydrogeologic test program with some subsequent modeling should provide answers to all of these questions, particularly during the design of a large potable water supply system. The following is a summary of the key components to a proper groundwater supply investigation:

- test drilling or exploration
- observing well construction in production zone
- testing initial water quality in production zone
- testing production well construction in production zone
- observing well construction in zone above and below production zone
- observing wells in production zone up-gradient and down-gradient of test site
- testing water quality in all affected aquifers
- testing aquifer performance on production zone to measure hydraulic characteristics
- additional aquifer testing on the production aquifer or adjacent aquifers, if necessary
- analyzing data and modeling
- determining safe yield

General rules govern the design of a wellfield located near a source of saline water or within an aquifer subject to vertical inflow of varying water quality. Whenever possible, a wellfield should be aligned perpendicular to the flow direction within the production aquifer. Drawdowns of water levels or pressures in individual wells should be minimized based on the hydraulic characteristics of the aquifer and the collapse strength of the well casing. Both of these problems can be evaluated with a computer model. If a screened well is required, then the well yield should be limited to a discharge rate that will maintain the entrance velocity of water through the well screens to less than 0.1 ft/s (see Driscoll 1986).

Membrane treatment facilities require one other significant design feature. The number of wells used to feed a single "train" of membranes should be a whole number. One well per train is best, but a larger number (two or more) can be used. At least one standby well for each five production wells should be constructed depending on the well yield and the size of the system.

Well construction and maintenance. Major problems at membrane treatment plants have been caused by improper or poor well construction. In many cases, well failure causes a sudden change in water quality. The most common causes of well failure are borehole collapse, corrosion of the casing, improper or defective construction techniques, growth of organisms within the wellbore, and formation of mineral concentrations or crusts in the open-hole or screened section of the wellbore.

Designing any water well involves a careful scientific evaluation of the aquifer to be developed, as previously discussed. Among the issues that must be considered in well design are drilling methodology, structural integrity, construction materials, and a determination of whether the well should be an open-hole type or screened. Most or all of the field information necessary to design production wells should be obtained during the hydrogeologic investigation.

If a well is used to provide source water for a membrane treatment facility, additional design factors must be considered. The sensitivity of the membrane to turbidity, oxidants, dissolved and particulate iron, and silica, as well as to rapid changes in the concentration of TDS must all be appraised in the individual production well design. The most complete references on well construction design are Driscoll (1986) and Campbell and Lehr (1973).

The process of designing large-scale municipal wellfields to feed membrane treatment facilities should follow a logical flow, from collecting hydrogeologic information to testing to finalizing design. The following is a generalized production well design procedure:

1. Drill a pilot hole.
2. Determine the production zone depth interval.
3. Evaluate production zone materials.
4. Determine the necessity for screening.
5. Select material.
6. Test the prototype well design production.
7. Test the aquifer.
8. Finalize production well design.
9. Observe well construction and field adjustments to design.
10. Test step-drawdown of all production wells.
11. Select pump type, pumping rates, column pipe, and valves.
12. Establish monitoring program.

During the design of a well or wellfield for a small membrane application, strictly adhering to this procedure may not be necessary; some steps may be deleted based on professional judgment.

The selection of well construction materials and the drilling method can significantly affect the future performance of a membrane treatment facility. Nonmetallic casing should be used whenever possible because membranes are sensitive to the concentration of iron and certain other metals in the water as well as to particulate material. Steel casings are particularly subject to corrosion when conveying saline water.

Polyvinyl chloride (PVC) can be used to case relatively shallow wells if structural considerations allow it. Before PVC is selected for certain applications, its structural properties should be assessed to avoid a casing collapse caused by the heat of hydration during grouting (National Water Well Association and the Plastics Pipe Institute 1980). Acrylonitrile-butadiene-styrene pipe has slightly better thermal properties but can be brittle at coupling points.

Fiberglass casings are recommended for applications where thermoplastic casings are structurally unstable. Some fiberglass casings are designed for use at depths ranging from 100 ft to several thousand feet. Fiberglass has been used to case deep industrial waste injection wells. The cost of fiberglass casing can become a significant factor at depths of over 1,000 ft.

In deep wells, casing strength can become a major issue. Stainless steel has been used to case deep water supply wells in the Middle East. The grade of stainless steel required is a design consideration; grade 304 stainless steel tends to be much more subject to corrosion in a groundwater system compared to grade 316 stainless steel. There are, however, examples of rapid corrosion and failure of grade 316 stainless-steel casings in deep wells. Using pipe to convey water of one salinity past water of a different salinity on the pipe exterior may cause cathodic corrosion, rapidly destroying nearly all types of metallic casing.

Another method of solving the structural casing problem is to construct a standard well with steel casing and then line the well with either PVC or fiberglass. This concept is most effective if the liner is cemented in place. In many cases, well screens must be constructed with grade 316 stainless steel because of strength and efficiency considerations.

Selection of the grout material is also a significant factor in the design of a well to be used for a membrane treatment facility. Cement grout can be chemically attacked and may fail in groundwater containing high concentrations of sulfate and various other ions. To increase the resistance to sulfate attack, particularly in deep supply wells, American Petroleum Institute (API) class B cement or American Society for Testing and Materials (ASTM) C 150, type II cement should be used whenever possible. The addition of bentonite gel to the grout slurry should be avoided in the lower part of the casing area adjacent to the open-hole section of the well because bentonite is a potentially significant membrane foulant. Since bentonite is commonly added to the cement grout slurry to reduce the temperature of the hydration heat and to reduce slurry movement into the formation, the use of PVC casing should be carefully evaluated. Cement grout should be pumped under pressure, with care taken not to damage the casing. In certain cases, the grout may have to be placed by tremie pipe or by a combination of methods.

During well construction, selection of the drilling method and type of drilling fluid can affect the operation of a membrane treatment facility. The use of standard bentonite drilling mud in the construction of the open-hole or screened section of the well is not recommended. Residual bentonite can foul the prefilters or membranes directly or, in the case of pretreatment with acid, can cause the formation of silica colloids that can foul the membranes. In permeable, lithified formations, the reverse air-drilling method can be used to drill the open hole. In the drilling of unconsolidated material to set screens, a decomposing polymer should be used. When an organic polymer fluid is being used, chlorinated drilling water prevents bacterial growth problems.

Conventional Treatment Versus Membrane Processes

In the development of a water supply for a conventional water treatment facility, source water entering the plant can vary widely in quality, particularly with regard to turbidity. As long as the concentration of dissolved solids or nonremovable ions, such as dissolved chloride, do not exceed drinking water standards, the water treatment facility will continue to function. For a membrane treatment facility, however, variation in water turbidity can result in fouling the prefilter system or membrane elements.

Membrane systems are also designed to treat water within specific ranges of dissolved solids and other constituent ion concentrations. If the source water contains concentrations of solids or ions outside the design limits, the membrane treatment facility may not be able to treat the water to potable standards or the plant may not function at the desired capacity. Therefore, the source water for a membrane treatment facility must be investigated so that future ranges in water quality fluctuations will be accurately assessed.

Development of a surface water source for a membrane treatment facility should be avoided unless a groundwater source is unavailable. Surface waters naturally have large fluctuations in both availability and quality. The intake structures constructed into surface water bodies are difficult to design and maintain. Most surface water intake systems must be modified to allow additional treatment before the source water can enter the membrane facility. Modified surface water intake structures include

- surface intake structures with a settling reservoir
- surface intakes with a canal recharge or wellfield system
- beach wells or galleries
- seabed filtration systems

If a modified surface water intake system is used, hydrogeologic investigations similar to those conducted for the development of a groundwater supply may be required.

The process of developing and maintaining a groundwater source for a membrane treatment facility is somewhat different than conventional wellfield development. As in the development of all wellfields, a hydrogeologic investigation should be conducted to select an aquifer, determine the number and location of production wells, assess the hydraulic properties and reaction of the selected aquifer to pumpage, and assess the quality of water in the aquifer and how it will vary in the future.

The major difference between groundwater supply development for a membrane process treatment facility compared to a conventional plant lies in the analysis of anticipated water quality changes and in well construction practices. For a membrane facility, detailed three-dimensional modeling of groundwater quality is necessary, particularly in the design of major treatment facilities. Individual production wells, pumps, and piping must be constructed with materials compatible with the membrane process. Nonmetallic materials should be used to avoid fouling the membranes with substances such as dissolved iron. Both the turbidity and silt density index (SDI) of the water produced from each well should be minimal because they exert some influence on the selected well yield and on the general design of the well.

In conclusion, the design, construction, and operation of a groundwater supply for a membrane treatment facility is slightly more complicated compared to a supply for a conventional treatment facility, but the differences are easily defined. However, some knowledge of the membrane process is necessary to make appropriate decisions during the design and construction process.

PRETREATMENT

Preserving membrane integrity through proper pretreatment of source water is essential to maximize the efficiency and longevity of an RO or NF membrane system. This section provides a more detailed discussion of the various methods of pretreatment that prevent chemical alteration of the membrane polymer and foulant accumulation and scaling on the membrane surface.

Precautions that can be taken to maintain suspended solids at an acceptable level in the membrane source water include a judicious selection of the water source (wells being more favorable than surface supplies), proper well construction, filtration, and, if necessary, the use of chemical pretreatments to assist in removing suspended solids. Modern high-performance polymeric antiscalants have proven highly effective in stabilizing supersaturated solutions of sparingly soluble salts and should be used whenever a supersaturated condition could occur. Severe biological fouling can generally be avoided by minimizing the time the plant is not operating to a day or less. Flushing the membranes with permeate water on shutdown is also helpful to minimize biological fouling. In more severe cases, a biocide may have to be used on either a continuous or intermittent basis. Care must be taken to remove residual free chlorine if the membrane is of a polyamide composition or to adjust the source water pH for a cellulosic membrane.

Chemical Alteration of Membranes

If an RO or NF membrane is chemically altered by substances in the source water, the membrane's efficiency will decrease. Chemical attacks on membranes generally occur via oxidation or hydrolysis, both of which are easily controlled. The susceptibility to chemical attack is a function of the chemical composition of the membrane polymer.

Cellulose acetate polymers are subject to a loss of rejection properties because of a gradual hydrolysis of the acetyl groups on the polysaccharide backbone. The rate of hydrolysis is a function of pH. Mineral acid is generally required to adjust the pH of most source waters into the 5.5 to 6 range to maximize membrane life. Polyamide membranes are capable of operating over a rather broad pH range: from pH 2 to 11 on a continuous basis and up to pH 12 for short-term cleaning. When a polyamide membrane is used for processing water at the higher pH (>8) range, careful consideration must be given to controlling calcium carbonate scaling, as discussed in greater detail in this section.

Although polyamide membranes are not particularly pH sensitive, all varieties of these membranes are to some extent intolerant of chlorine and certain other oxidizing agents. Adding sodium metabisulfite to the source water in excess of what is required to reduce the chlorine should protect the more tolerant polyamide membranes from chlorine attack. For the most susceptible polyamide compositions, chlorine should be totally avoided as a part of pretreatment. Cellulosic membranes, on the other hand, will tolerate chlorine at levels commonly used to provide disinfection of potable water supplies, 1 mg/L Cl_2 or less. However, over long membrane service lives, the chlorine will facilitate oxidation of the membrane material.

Membrane Fouling

An accumulation of one or more foreign substances on the surface of a membrane will result in a loss of flux. Higher operating pressures will then be required to maintain water production and quality. Membrane fouling generally occurs by one of the following mechanisms:

- deposition of silt or other suspended solids that were inadequately removed by the pretreatment
- inorganic scale deposits formed because of a concentration of sparingly soluble salts in excess of their solubility limit
- biological fouling caused by excessive microbial growth
- interaction of organics with the membrane

Silt and Colloidal Fouling

Membrane manufacturers usually require one or more of the following guidelines for suspended solids

- turbidity less than 0.2 ntu
- zeta potential of more than –30 mV
- maximum silt density index (SDI) of 3 to 5

Turbidity is best measured by an in-line continuous monitor. Zeta potential, used for fresh and brackish water, measures a colloid's electrical stability with a meter that can also be installed for continuous monitoring. Silt density index, as described in ASTM method D4189, is based on the rate of plugging a standard 0.45-micron membrane filter. The procedure is shown in appendix D.

Silt density index values of 4 for hollow fiber membranes and 5 for spiral-wound membranes are approximate upper limits recommended by various manufacturers. If these levels are exceeded, pretreatment is definitely required; at lower levels, pretreatment may be recommended to reduce cleaning cycles. The minimum pretreatment for suspended solids, regardless of the SDI, is a 5-micron final filter just before the high-pressure pump. If the SDI exceeds the manufacturer's recommendations, it may be possible to achieve sufficient reductions by other filtration techniques listed in Table 2-15.

For those cases where contaminant particle size is so small that filtration alone is not sufficient, adding chemicals to coagulate the colloidal particles is necessary. These chemicals destabilize the colloidal particles by neutralizing the electrostatic charge that surrounds the particles. Once neutralized, the particles are readily agglomerated and removed by filtration.

The use of coagulating chemicals for pretreating RO or NF source water must be approached with caution. Properly applied, these chemicals are very effective. When overdosed or dosed at the wrong pH, their use can cause more problems than it solves. It is not possible to predict the feed rate based on chemical or physical parameters of the source water. Jar testing or pilot plants are essential to determine the optimal chemical treatment and level of application. With surface waters, the level of suspended solids can vary widely depending on a host of environmental factors. This necessitates frequent checks to determine the proper feed rate to maintain the required SDI. On-line instruments, such as streaming current detectors, can be applied to control chemical dosing rates.

Common coagulant–flocculant chemicals are listed in Table 2-16. Of those listed, the least likely to react with the membrane are the anionic or nonionic polymers. Ferric and aluminum salts can be used effectively provided the pH is properly controlled, sufficient time is allowed to fully precipitate the metallic hydroxide, and the precipitates are completely removed by the filter. Otherwise, the membrane can be fouled by aluminum or iron hydroxides. In case of an upset, both of these inorganic coagulants can be removed from the membrane by

Table 2-15 Pretreatment techniques (listed in decreasing order of effectiveness) for reducing silt densities

Method	Comments
Ultrafiltration	Relatively expensive. Frequent backflushing and periodic cleaning are required. Operates at slightly higher pressures than microfilters. Removes higher-molecular-weight organics (>100 kD) in addition to particulates.
Microfiltration	Operates at low pressures and requires periodic chemical cleaning and frequent backwashing. Removes particulates but no dissolved organics. Cost may be comparable to coagulation–flocculation–clarification and direct filtration for small systems.
Precoat filtration	Also relatively expensive. Precoat media must be added to replace losses in backwash step. Does not remove dissolved organics.
Multimedia and pressure filtration	Best alternative if workable. Pilot testing may be required to determine the best media for SDI reduction. May be used in conjunction with coagulants–flocculants, with and without settling.
Rapid sand filters and ultrahigh-rate filters	Coagulants are generally required to achieve acceptable SDI reduction, most typically through in-line coagulation. Can handle limited turbidity levels.

Table 2-16 Common coagulant-flocculants

Chemical	Comments
Inorganics	
Aluminum sulfate or alum, $Al_2(SO_4)_3$	Minimum solubility is achieved at pH 5.7 to 7.8. Must allow sufficient time for complete precipitation or aluminum salts may deposit on the membrane.
$FeSO_4$	Oxidation to ferric form of iron is required to be effective. Optimal pH is >8.4.
$Fe_2(SO_4)_3$	Works best at high pH.
$FeCl_3$	Works best at pH 6.0 to 7.0. Chemical is widely used..
Organic Polyelectrolytes	
Cationics	Will interact with antiscalants. Can irreversibly foul many spiral-wound membranes.
Anionics and nonionics	Will not react with antiscalants. May foul certain membranes. Consult manufacturer. Generally less effective than cationics as coagulant–flocculants.

cleaning, with iron being somewhat easier to dissolve than aluminum. Cationic polyelectrolytes are excellent coagulants but, if overdosed, will foul membranes rapidly and sometimes irreversibly. Cationic polymers can also react with anionic scale inhibitors to form insoluble organic deposits that can be very difficult to remove by cleaning.

Although the cautionary information in the preceding discussion is important, silt or colloidal fouling is not a widespread problem with groundwater sources. Most plants drawing water from brackish aquifers or hard groundwaters in Florida operate successfully using only a 5-micron filter for protection against particulate fouling. Certain large seawater RO plants have been able to avoid pretreatment for colloidal fouling problems by drawing the water from a beach well rather than from a surface intake.

Scale with crystalline morphology. Scale forms on a membrane surface when naturally occurring, sparingly soluble salts are concentrated beyond their solubility limit by the membrane process. Table 2-17 lists the commonly encountered scales and the methods for preventing precipitation on a membrane. The methods will be discussed in greater detail later.

The solubility limit of a sparingly soluble salt is defined by the salt's solubility product constant K_{sp}, which is discussed in more detail in appendix C. Membrane manufacturers have design software that performs complex calculations to yield values of K_{sp}. For ease of discussion and preliminary analysis, most chemists and engineers prefer to simplify the mathematics by defining a conditional solubility product constant (K_{sp}) that specifies solubility as a function of ionic strength (I). The ionic strength of a solution may be defined as follows:

$$I = \frac{1}{2}\sum m_i z_i^2 = \frac{1}{2}(m_1 z_1^2 + m_2 z_2^2 + \dots + m_n z_n^2) \quad (2\text{-}1)$$

Where:

m_i = molar concentration of the ith species
z_i = ionic charge of the ith species
n = total number of ionic species in the solution

Table 2-17 Common scales and other inorganic foulants and methods for controlling them

	Method of Control	
Species	Primary	Secondary
Calcium carbonate, $CaCO_3$	Mineral acid Threshold inhibitor	Soften
Calcium sulfate, $CaSO_4$	Threshold inhibitor Soften	Limit recovery
Barium sulfate, $BaSO_4$	Threshold inhibitor Soften	Limit recovery
Strontium sulfate, $SrSO_4$	Threshold inhibitor Soften	Limit recovery
Calcium fluoride, CaF_2	Threshold inhibitor Soften	Limit recovery
Silica, SiO_2	Limit recovery	–
Ferric hydroxide, $Fe(OH)_3$	Precipitation and filtration	Soften
Aluminum hydroxide, $Al(OH)_3$	Precipitation and filtration	–
Sulfur, S	Exclude air and other oxidants	–

An example of an ionic strength calculation is given in Table 2-18. In addition, Figures 2-4 and 2-5 illustrate how the solubility of the sulfate salts of calcium, barium, and strontium vary with the ionic strength of the solution.

Whenever the ratio of the ion product to the K'_{sp} exceeds 1.0, there is a potential for scale deposition on the RO or NF membrane and pretreatment is essential. (Calculation of an ion product is also discussed in appendix C.) For spiral-wound membranes, because of concentration polarization at the membrane surface and occasional water quality variation, pretreatment is recommended whenever the ratio of the ion product to K'_{sp} is greater than 0.8.

In theory, supersaturation of all insoluble salts can be evaluated in this manner. However, when examining calcium carbonate solubility, engineers prefer to evaluate scaling potential in terms of either the Langelier Saturation index (LSI) or, in the case of seawater, the Stiff and Davis scaling index (SDSI). Both indexes are explained further in appendix E. For either index, a positive value indicates a scale-forming condition. For most brackish waters, the LSI has proven to be an adequate predictor. For seawater, which has a much higher ionic strength, the SDSI has become the preferred index. Nomographs have been widely published to allow one to calculate either index without having access to logarithm tables or a calculator that has scientific functions. Membrane manufacturers, consulting engineers, and technical service departments of chemical suppliers have computer programs that can either be made available or accessed on request to evaluate the supersaturation of sparingly soluble salts at various levels of recovery from a source water of known composition. Results from one such program are given in Figure 2-6.

The following are some methods for preventing scale (as briefly mentioned in Table 2-17):

- acidify to remove carbonate or bicarbonate ions
- limit recovery to prevent salts from concentrating in excess of solubility limits

Table 2-18 Example of an ionic strength calculation

Species	Assumed Concentration, *mg/L*	Number of Milligrams per Mole	m_i*	z_i^2	$m_i z_i^3$
Calcium, Ca^{2+}	409	40,000	0.0102	4	0.0408
Magnesium, Mg^{2+}	169	24,300	0.00695	4	0.0278
Sodium, Na^{+}	314	23,000	0.0137	1	0.0137
Bicarbonate, HCO_3^{-}	150	61,000	0.00246	1	0.00246
Sulfate, SO_4^{2-}	1,510	96,100	0.0157	4	0.0628
Chloride, Cl^{-}	500	35,500	0.0141	1	0.0141
				$\sum m_i z_i^2$	= 0.1617
				$I = \frac{1}{2}\sum m_i z_i^2$	= 0.0808

*Determined by dividing the value in column 2 by the value in column 3.

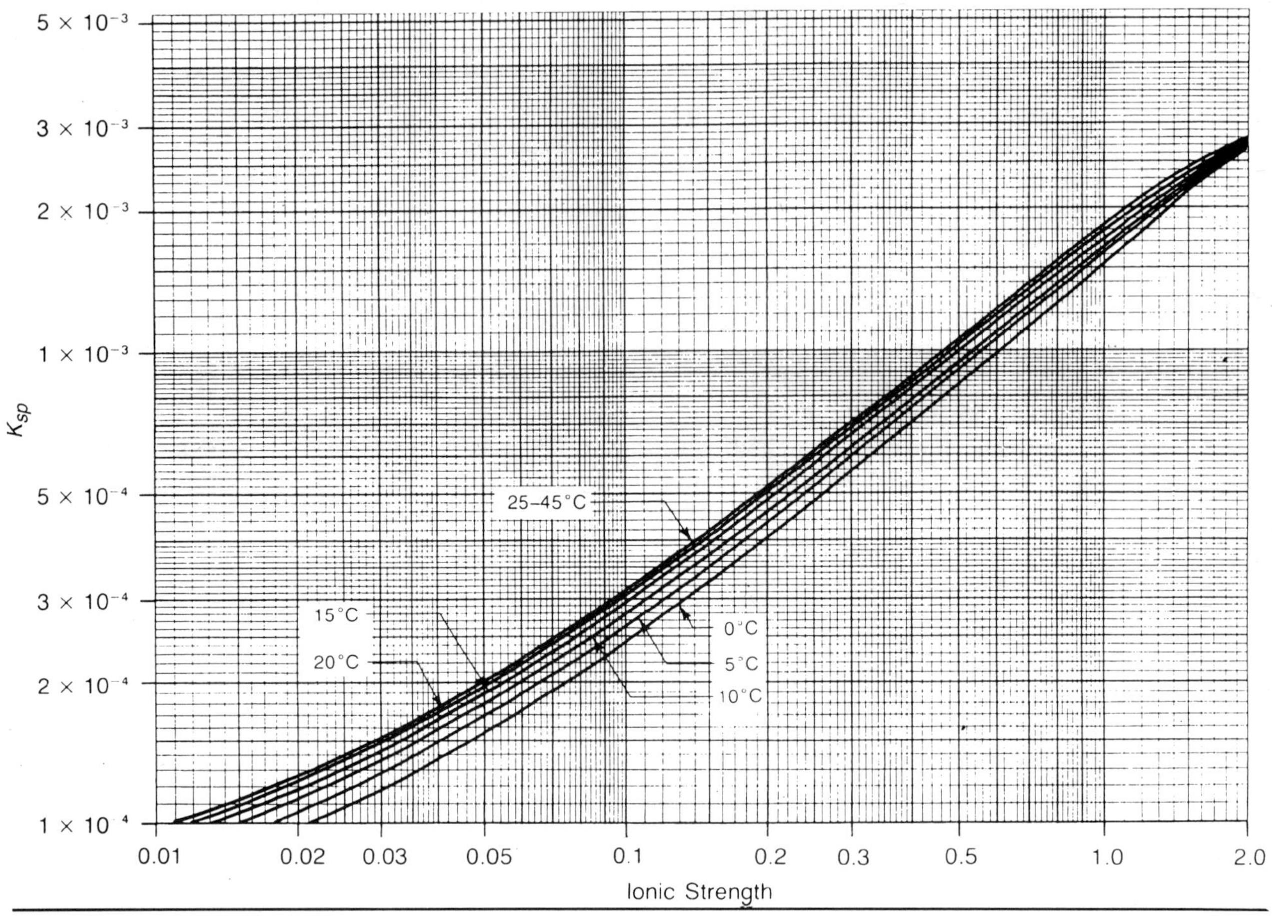

Figure 2-4 Calcium sulfate solubility versus ionic strength

- soften the source water to remove calcium ions
- use threshold antiscalants, which are effective for calcium sulfate and calcium carbonate

Acidification is effective in preventing calcium carbonate and magnesium scaling, provides a small benefit in terms of increased silica solubility, and is usually required to prevent hydrolysis of cellulose acetate membranes. The obvious safety hazards associated with handling strong mineral acids necessitate appropriate operator training and conscientious attention to safe handling practices. Cation exchange softening is impractical on most brackish waters because of the high levels of hardness in the water; it is not feasible with seawater. Lime softening is practiced occasionally for brackish waters with exceptionally high hardness, but the costs of capital, labor, chemicals, and sludge disposal are high. Limiting recovery to ensure solubility limits are not exceeded should be viewed as a last-resort method for dealing with crystalline scales because the practice wastes a substantial portion of the groundwater resource. The addition of a threshold antiscalant is the most practical and widely used method to prevent crystalline scale from forming on membranes.

Threshold antiscalants have often been incorrectly referred to in desalination literature as *sequestrants*. These chemicals do not sequester by being in exact proportions for interaction with calcium, barium, or strontium. Instead, they react

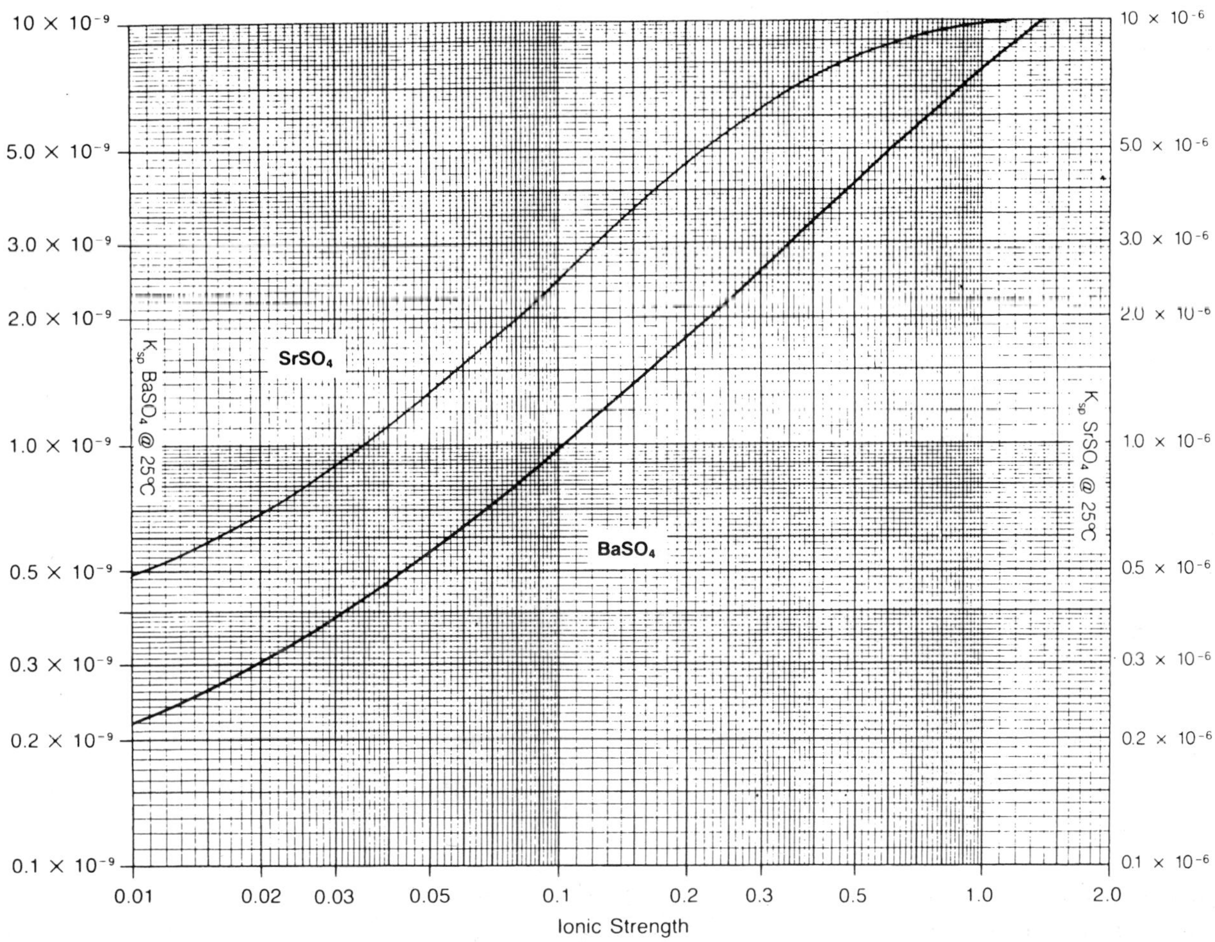

Figure 2-5 Barium sulfate and strontium sulfate solubility versus ionic strength

with the surfaces of ion clusters to prevent the orderly growth needed to form the scale crystal or allow it to grow to a harmful size. Because of this unique property, antiscalants can be added at very low levels, usually less than 10 mg/L, into the membrane source water to prevent scale from forming at multiple levels of supersaturation. Two chemical antiscalants that have achieved widespread usage in the production of potable water are polyphosphates and high-performance polymeric antiscalants, most typically polyacrylates.

Before the development of polymeric antiscalants, sodium hexametaphosphate (SHMP) was widely used in membrane systems to prevent sulfate scales from precipitating in a membrane system. However, SHMP has several potential drawbacks:

- It is less effective than the new polymeric antiscalants, so higher doses of SHMP are normally required.
- The solid form of SHMP has to be diluted for injection. Once diluted in water, it starts to hydrolyze to the orthophosphate, which is ineffective as a scale inhibitor and contributes to calcium phosphate scaling.

RO system	Plant Name
Percent recovery	80

Feedwater Analysis

Cations	mg/L	Anions	mg/L
Ca	100	HCO_3	189.9
Mg	140	SO_4	350
Na	870	Cl	1,600
K	40	F	1.6
Ba	0.1	NO_3	0
Sr	20	PO_4	0
Fe	0	CO_3	0.28
Silica	14	CO_2	13.53
Temperature	25°C	Pressure	300 psi

Parameter	Value
Ionic strength, feedwater	6.800001E-02
Total cation as $CaCO_3$	2,743 mg/L
Total anion as $CaCO_3$	2,780 mg/L
Total dissolved solids, feedwater	3,096 mg/L
pH, feedwater	7.2
Total dissolved solids, concentrate	15,479 mg/L
pH, concentrate	7.91

Polymeric Antiscalant Dosage*

Constituent	Minimum Dose, mg/L	Recommended Dose, mg/L
$CaCO_3$	3	5
$SrSO_4$	2	4
$BaSO_4$	3	5
CaF_2	–	5

Degree of Saturation of Potential Scales

$CaCO_3$	1.94%	$SrSO_4$	250%
$CaSO_4$	29%	$BaSO_4$	2,375%

Ion Product

Constituent	Ionic Strength
$CaCO_3$	2.903823E-07
$SrSO_4$	2.078944E-05
CaF_2	2.211644E-09
$CaSO_4$	2.272678E-04
$BaSO_4$	6.630918E-08

* To ensure scale-free operation, polymeric antiscalant should be dosed consistent with the greatest scale potential in the system as indicated by the highest dosage given.

Figure 2-6 Sample computerized evaluation of the supersaturation of sparingly soluble salts for a given source water, with a polymeric antiscalant dose projection

- Discharging the concentrate stream, which contains phosphates, can contribute to environmental problems.
- It can contribute to biological growth within the membrane element.

High-performance polymeric antiscalants were developed to offset the deficiencies attributable to SHMP and to provide a dependable, cost-effective pretreatment alternative. Operating data found in Figures 2-7 and 2-8 demonstrate the scale control efficacy of a polymeric inhibitor.

The data in Figure 2-7 are from a pilot test conducted at the Island Water Association in Sanibel, Fla. Operating without mineral acid to reduce the feed pH, a 3-mg/L dose of polymer provided effective control of calcium carbonate scaling at an LSI of +2 in the concentrate stream. The concentrate stream was also supersaturated in strontium and barium sulfates. After the test had concluded, the utility switched the entire plant to the polymeric antiscalant to save the expense of adding acid. The plant continues to operate without any scaling problems.

Figure 2-8 presents data collected from a pilot RO system operating on a highly supersaturated calcium sulfate water at Venice, Fla. Again, the stability of the normalized data indicates effective inhibition of the sulfate scales of calcium, barium, and strontium, as well as of calcium carbonate. Because this plant was constructed in two sections and about half of the membranes are cellulosic, it continues to operate with a combination of sulfuric acid and polymeric antiscalant.

Threshold antiscalants are usually added to a membrane system ahead of the final cartridge filter. If media filtration or activated carbon are used, antiscalants should be added after these steps. Antiscalants are normally pumped either from a dosing tank or directly from the drum into the source water at a rate that is determined by the size of the RO or NF system, the recommended dose, and the delivery range of the dosing pump. It is best to regulate the dosing pump delivery by adjusting the stroke length while keeping the stroke frequency as high as possible to achieve an even distribution of chemical.

The correct dosage is most easily and accurately verified by using a stopwatch to note the time to draw down a small volume from a calibration tube. A permanent arrangement whereby a graduated column is permanently fixed to the suction side of the dosing pump and valved so that the antiscalant solution can be drawn from the graduated cylinder, rather than from the main dosing tank, provides a convenient method for periodically checking the antiscalant dose. The following formula relates the fluid drop from a graduated cylinder to the dosage:

$$FD = \frac{AD \times EF}{AC} \times 0.0316 \qquad (2\text{-}2)$$

Where:

FD	=	fluid drop, in mL/min
AD	=	antiscalant dosage, in mg/L
FF	=	feedwater flow, in gal/min
AC	=	antiscalant concentration, in lb/gal
0.0316	=	conversion factor to account for use of different systems of units

Location: Sanibel, Fla.
RO Feed: 3,700 mg/L TDS

Membrane: Thin-film composite spiral wound
Recovery: 80 percent

Scaling Potential:

LSI:	+2.0
Strontium Sulfate:	156 percent of Saturation
Barium Sulfate:	226 percent of Saturation

Antiscalant Dose: 3 mg/L

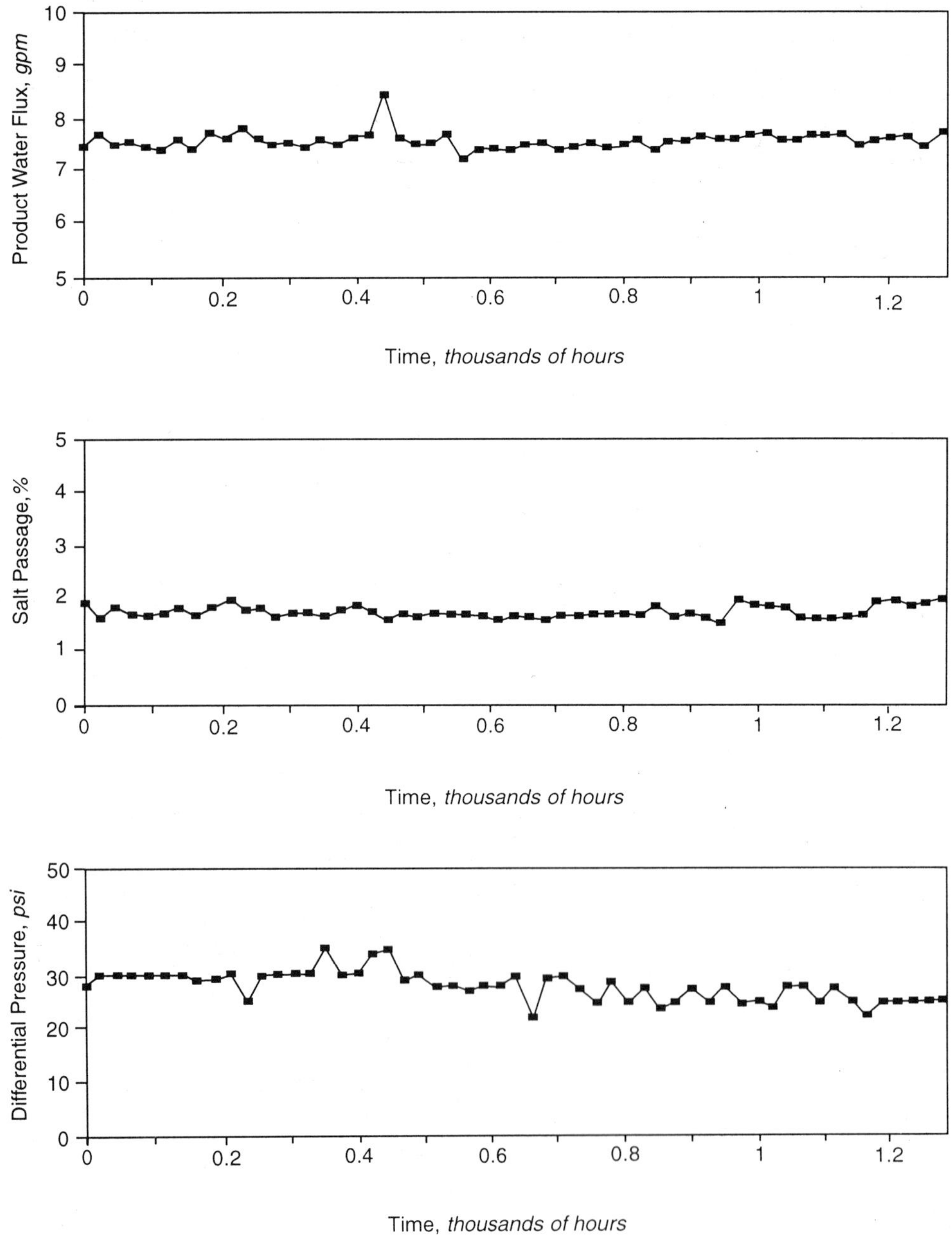

Figure 2-7 Scale inhibition study, Sanibel, Fla.

Location: Venice, Fla.
RO feed: 2,600 mg/L TDS

Membrane: Hollow Fiber
Recovery: 57 percent

Scaling Potential:

LSI:	+1.3
Calcium Sulfate:	157 percent of Saturation
Strontium Sulfate:	271 percent of Saturation
Barium Sulfate:	1,816 percent of Saturation

Antiscalant Dose: 5 mg/L

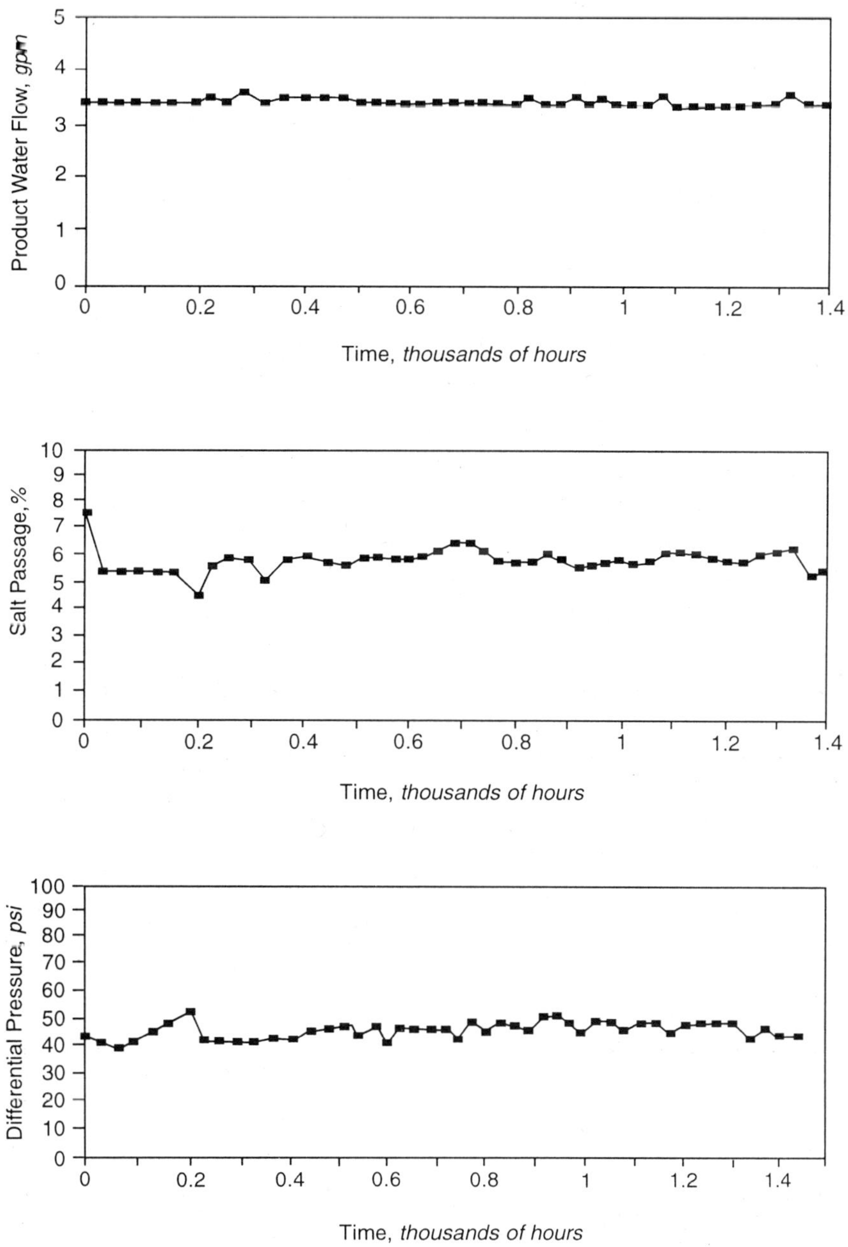

Figure 2-8 Scale inhibition study, Venice, Fla.

Amorphous deposits. Many groundwaters contain hydrogen sulfide, which can be oxidized to elemental sulfur if air or other oxidants are introduced into the membrane system. If such an air leak occurs before the final filters, a certain portion of the colloidal sulfur may be caught by the filters and membrane damage can be minimized. However, a significant proportion of the colloidal sulfur is smaller than the 5- to 25-micron effective retention size of the filters and will pass through to the membrane elements. It is especially important to ensure that air is not allowed to enter the module from the reject or product side of the system on shutdown. If it does, colloidal sulfur fouling, which is difficult to remove by cleaning, will be inevitable. No chemical pretreatment can prevent colloidal sulfur fouling; good system design is the best prevention.

Silica is another example of a substance that, if its solubility is exceeded,.can form an amorphous, inorganic deposit on a membrane. Because silica is a polymer rather than having a crystalline ionic lattice, threshold scale inhibitors (antiscalants) have not been effective in stabilizing supersaturated silica solutions. Currently, the only realistic way of preventing silica deposition is to limit recovery. Nonetheless, the slow kinetics of silica polymerization have allowed some well-designed membrane systems to operate at the theoretical solubility. Figure 2-9 shows the theoretical silica solubility in the pH range of 7 to 7.7 as a function of water temperature; the values

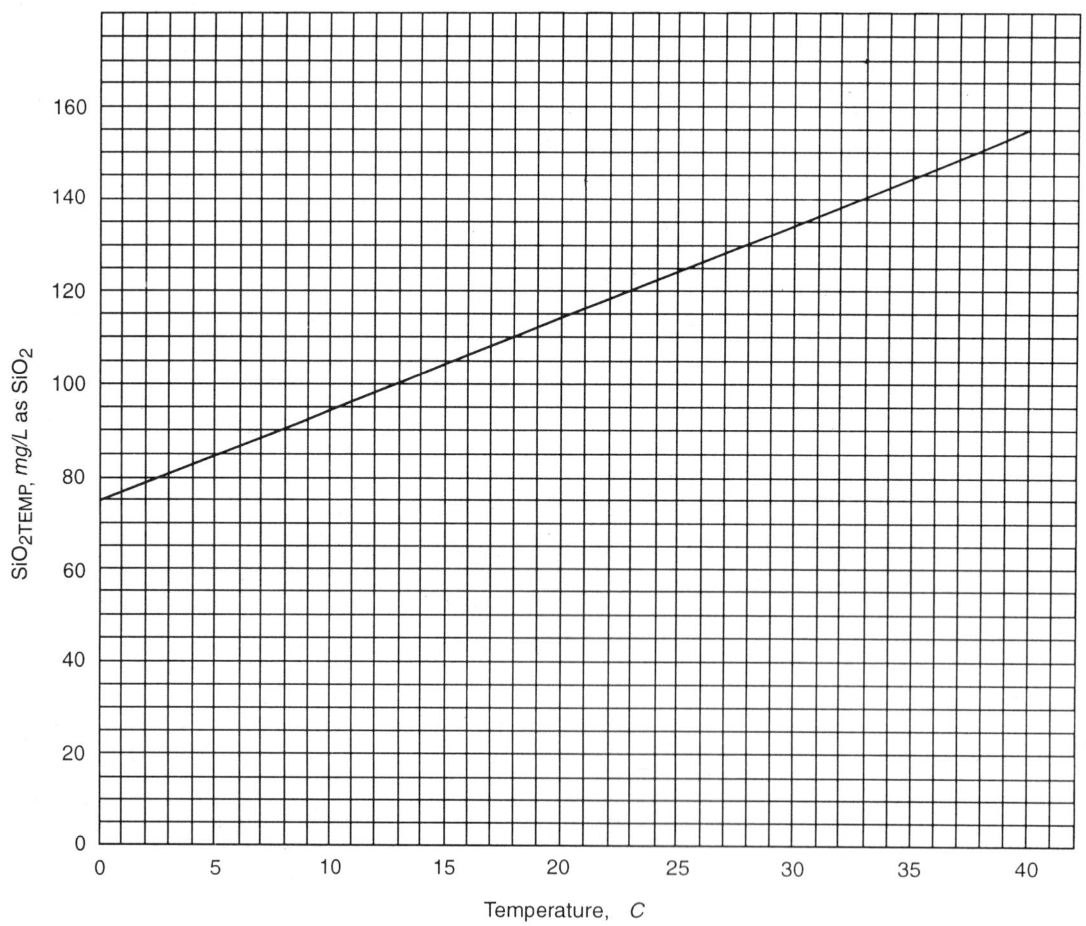

Figure 2-9 Effect of temperature on theoretical SiO_2 solubility, pH 7.7

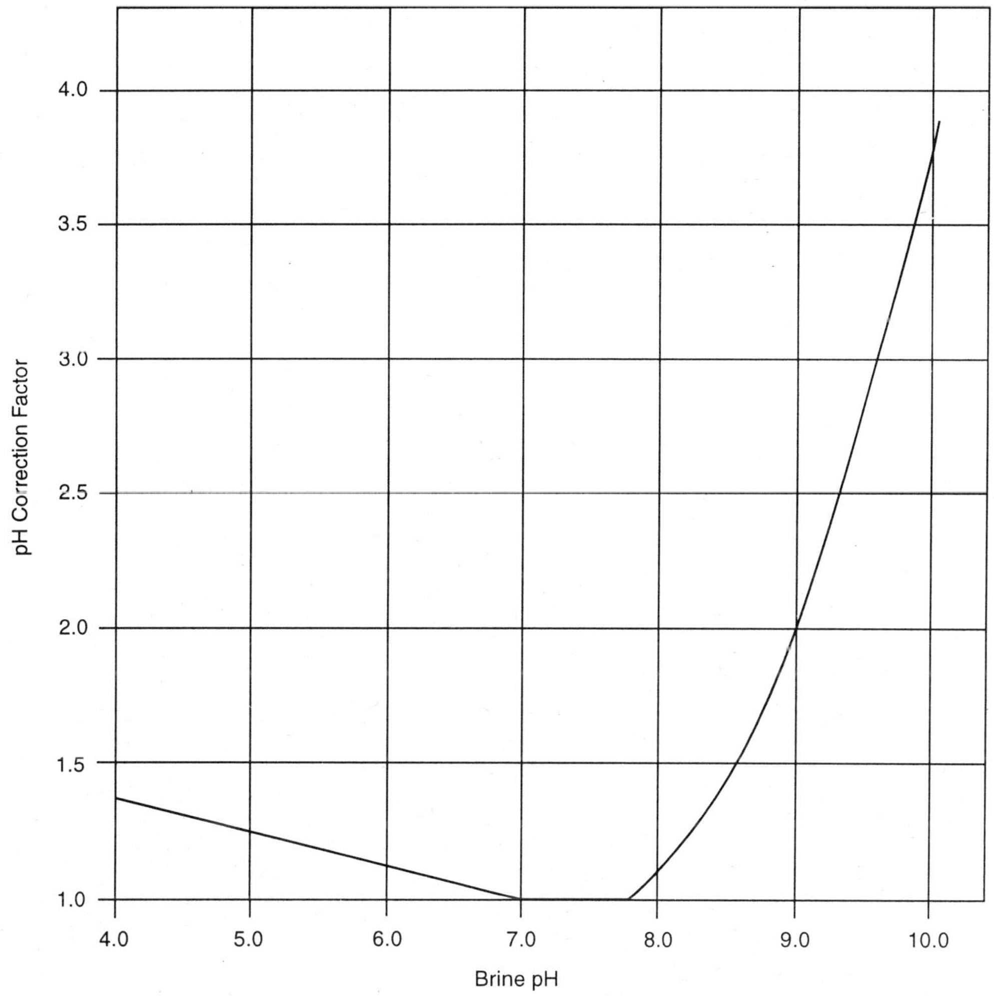

Figure 2-10 SiO_2 pH correction factors

in this figure also indicate maximum levels of silica in the brine stream that have been achieved without precipitation on the membrane.

Silica solubility can be altered by pH adjustment as shown in Figure 2-10. The effect is less dramatic at low pHs than at the high end of the pH scale.

Another amorphous foulant occasionally appearing in membrane plants is iron hydroxide. When found in groundwater, iron is usually in the form of soluble ferrous bicarbonate. Provided the system is maintained in an anaerobic condition, ferrous iron is unlikely to cause problems. However, if allowed to oxidize, ferric iron will precipitate as the hydroxide in the pH range in which most waters are processed. Some membrane manufacturers specify maximum levels of iron permissible in the source water.

If removal is needed to meet operating guidelines under the manufacturer's warranty, iron is usually removed by oxidation followed by filtration. Manganese greensand filtration is a convenient way of doing both in one step (Figure 2-11). Care should be taken to prevent excess oxidant, such as potassium permanganate, from contacting membranes. Air oxidation or chlorine addition followed by filtering through appropriate filter media is another option, provided the same precautions are taken to minimize residual iron as when used as a coagulant–flocculant. Feedwater with a free chlorine residual should not contact thin-film composite

Figure 2-11 Greensand horizontal filters (from a 9.6-mgd membrane softening plant, Dunedin, Fla.)

membranes. For source water with high ferrous and low ferric iron levels, precipitates can be held in suspension by lowering the pH to 6.0 or less and adding dispersants and chelants. Microfiltration and ultrafiltration (UF) have also been effectively used to remove undissolved iron, and iron dispersants are now included with polymeric scale inhibitors.

Iron fouling may also occur by deposition of corrosion products from mild steel or cast-iron pipe in the system. Such fouling is generally found in older, often smaller systems where existing materials have not been replaced. Proper selection of construction materials is the best insurance against this type of iron fouling.

Precipitation of other metal hydroxides is rare with groundwater sources but occurs occasionally when alum-pretreated surface water is processed by membranes. No matter what the source of the alum, if a source water is treated with alum for removal of suspended solids before membrane processing, precautions must be taken with regard to pH and reaction time to ensure as complete a precipitation of the aluminum hydroxide as possible. Filtration systems must be maintained in optimal condition to prevent aluminum precipitates from getting into the membrane feedwater; otherwise, membrane fouling is likely to occur.

Biological fouling. Control of biological fouling is essential if a membrane system is to perform at design levels on a consistent basis. Because the interior of a

Table 2-19 Pretreatment for control of biological activity*

Treatment	Comments
Chlorination	Dechlorination required for polyamide membranes.
Copper sulfate	Effective for algae. Generally limited to seawater systems.
Chloramines	Some polyamide membranes not affected, but these chemicals are not as efficient as other disinfectants. Long contact time may be required.
Shock Treatment	**Comments**
Sodium bisulfite	Minimum dose of 500 mg/L for at least 30 minutes. Time between treatments depends on bacteria counts. Will not affect most membranes.
Iodine	Effective, but life of polyamide membrane will be shortened. Maximum dose is 15 mg/L.
Hydrogen peroxide	Same as for iodine. Presence of metals on membrane surface should be minimized during use.

*Biological control methods should be approved by the membrane manufacturer in order not to void membrane warranties.

membrane module or permeator is damp, it is an ideal place for microorganisms to grow. Once established on the membrane, microbes can adhere to the polymer tenaciously. More harmful than the organisms themselves is the biofilm they secrete. This biofilm uniformly covers the membrane, causing a deterioration in flux and salt rejection (SR). Fortunately, the population of microorganisms in most brackish ground-water sources is small enough that pretreatment for biological fouling is not required, provided the membrane system is used on a fairly continuous basis.

One of the best steps toward eliminating biological fouling is to size the membrane system capacity to meet the need for water. Any membrane that is allowed to sit out of service for more than a day or two should be flushed with pretreated water on a daily basis to prevent a tolerably small bacteria population from growing to an intolerable level. If the system is to be out of service for extended periods of time, it is necessary to flush the membranes with permeate and add a sanitizer recommended by the manufacturer. Formaldehyde and sodium bisulfite solutions are effective in preventing microbes from proliferating in the membrane modules during storage, but the use of formaldehyde is increasingly restricted because of its possible carcinogenic properties. The chemical gluderaldehyde is increasingly being used as a biocide during membrane storage.

Surface waters are more likely to need some type of pretreatment to prevent biofouling. For the most highly contaminated waters, continuous chlorination is necessary for membranes that can tolerate free chlorine. For less severe problems, periodic shock treatments with high chlorine doses may be the only treatment required to maintain an acceptable bacteria count. Table 2-19 summarizes current methods used to control biological fouling in membrane systems. Only products registered with the US Environmental Protection Agency (USEPA) and labeled for such use should be considered.

Membranes used to process tertiary sewage effluent (the most notable being Water Factory 21 in the Orange County, Calif., Water District) are noted for having severe but manageable biofouling problems. It has been demonstrated that bacterial adhesion to a membrane does not require a high population of viable organisms. Even if inactivated by chlorine, certain species of bacteria can adhere and provide a substrate for successive growth of other bacterial types on top of the original layer of microbes (Ridgway et al. 1984; Ridgway, Rigby, and Argo 1985). In such severe cases, frequent cleaning cycles are necessary to restore product flow.

Biological degradation. Theoretically, cellulose membranes can be biologically degraded by cellulose-producing bacteria. Consequently, feedwaters to cellulose membranes are typically chlorinated. Documented biological degradation has rarely been reported.

Organic fouling. The predominant cause of organic fouling is overdosing of high-molecular-weight cationic polymers used for coagulation. Oils, grease, and other water-immiscible organics must also be kept away from the membranes or else irreversible fouling can occur. Groundwaters rarely contain such foulants, but careful selection and application of activated carbon prefiltration is advised if these foulants are present. High-molecular-weight, weakly charged, dissolved organics—such as the humic and fulvic acids naturally found in many water supplies—can foul RO and NF membranes alone or in combination with other feedwater contaminants.

PROCESS TECHNOLOGY

Water-permeable membranes used in RO and NF are classified according to type, material used, and configuration or packaging. The two basic types of membranes currently used are asymmetric homogenous and composite. A wide variety of materials are used, including cellulose acetate and such polymers as aromatic polyamides. Two prominent configurations are sheet membrane in a spiral-wound device and hollow fiber in a U-tube device.

The first commercial membranes were asymmetric (nonuniform density) brackish water RO membranes with a thin, dense surface, 0.1–1.0 micron thick on an otherwise spongy, porous film 80–100 microns thick (Figure 2-12). The surface or skin gives the membrane its rejection characteristics, and the porous substrate supports this skin to withstand pressure differences of 1,200 psi in some seawater RO applications.

The first material for the asymmetrical-type RO membranes was cellulose acetate, with the dense skin obtained through an annealing process. Modified cellulosic membranes, including cellulose diacetate and cellulose triacetate, are still being used today because they are less expensive than noncellulosic membranes. Cellulosic membranes are produced in both sheet and hollow fiber form and generally have a rated water flux of approximately 12 gpd/ft^2 at standard conditions (400 psi, 25°C, and 2,000 mg/L NaCl).

Membranes made from the cellulosic materials have some major limitations. Chiefly, cellulose acetate hydrolizes back to cellulose over a period of time, resulting in a loss of SR even under ideal conditions. Figure 2-13 indicates that hydrolysis of cellulose acetate is both pH and temperature influenced. Hydrolysis is accelerated at warmer water temperatures. The optimal feedwater pH is approximately 5 for cellulosic membranes, and the pH must be maintained between 4.5 and 6.5 to maximize the membrane's useful life. Chemical cleaning must be performed within a relatively narrow pH range (4 to 8), which can be a disadvantage in terms of creating conditions that promote fouling. Cellulosic membranes are also subject to deterioration in the presence of microorganisms capable of cellulose enzyme production. However, because of the material's substantial oxidant tolerance, this risk can be avoided by source water chlorination to 1 mg/L.

Aromatic polyamide (arramed) membranes can operate in a pH range from 4 to 11 without any hydrolysis and are not susceptible to biological attack. Because the material is sensitive to oxidants, chlorinated water supplies must be dechlorinated before contact with the membrane. Oxidants degrade membranes, resulting in a loss

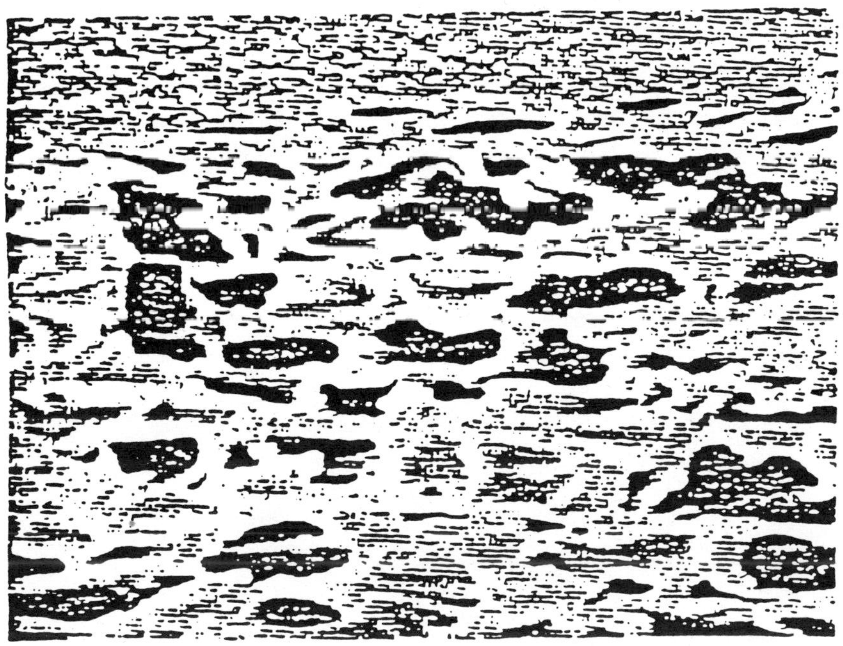

Figure 2-12 Cross section of an asymmetric membrane

of SR. This material was initially produced in the form of a hollow fine fiber 85 microns in diameter with a 42-micron bore, having a rated water flux of approximately 3 gpd/ft^2 under standard conditions. Aromatic polyamide membranes are also available in sheet form.

Asymmetrical membranes are subject to compaction. Compaction occurs mainly in the porous spongy substrate and results in the loss of productivity (water flux) over a period of time. Although it is pronounced in high-pressure seawater RO, compaction also occurs in brackish applications operating at one third to one fourth the seawater RO pressures.

The most recently developed types of membranes are composites. Composite membranes are formed by interfacial polymerization on a separate microporous layer, usually polysulfone, which replaces the conventional substrate to support the ultrathin (250–2,000 Å) salt barrier, as shown in Figure 2-14. A variety of polymers are used for the barrier layer, including the most common fully aromatic polyamides: aryl-alkyl polyamids and polyurea. The initial thin-film composites used aryl-alkyl polyamides as the salt barrier and were extremely sensitive to oxidants, which adversely affected long-term performance. This problem was modified by the introduction of polyurea, which is less sensitive but still has essentially no chlorine tolerance. A common feature of these polymers is their slightly cationic character, which also makes them adversely sensitive to anionic coagulants for pretreatment as well as to surfactants or detergents. Composites are also made with a salt barrier of

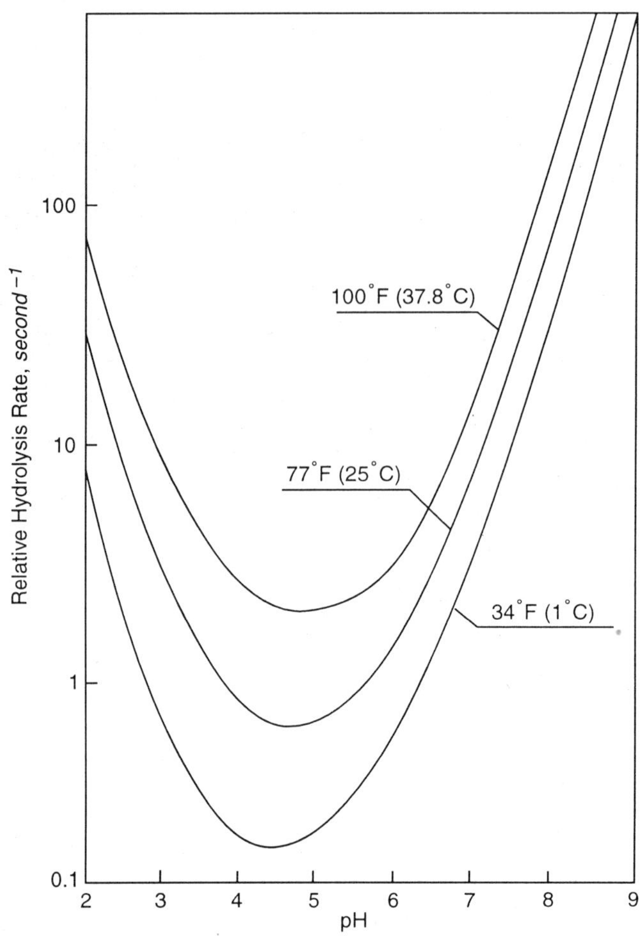

Figure 2-13 Effect of temperature and pH on hydrolysis of cellulose acetate

fully aromatic cross-linked polyamides that feature more of the desirable characteristics than any other membrane to date.

Figure 2-15 illustrates pH and temperature parameters for several different membranes. Composite membranes can operate over a pH range from 3 to 12, are not susceptible to hydrolysis or biological attack, and have some limited tolerance to oxidants. However, chlorinated supplies are still dechlorinated before contacting the membrane to enhance long-term performance. Since these membranes carry a slightly anionic charge, they exhibit no adverse reaction to anionic polyelectrolytes or cleaners. Composite membranes are not sensitive to selected cationic polyelectrolytes, but, in contrast to arramed hollow fiber membranes, they experience a severe loss of flux if cationic cleaners are used. Typical water flux ratings are 16 gpd/ft^2 at 200–300 psi for brackish water RO membranes.

Nanofiltration employs another type of composite membrane using modified polyamides and piperazinamides for the barrier layer. Such membranes, referred to as nanofilters, have high rejection (98 percent) for selected salts, such as magnesium sulfate, and reduced rejection for others, such as sodium chloride. These filters allow for cost-effective water-softening applications because the operating pressures of 70 to 125 psi are substantially lower than conventional RO operating pressures of 200 to 400 psi. Nanofilters also reject over 90 percent of most dissolved minerals. Because of

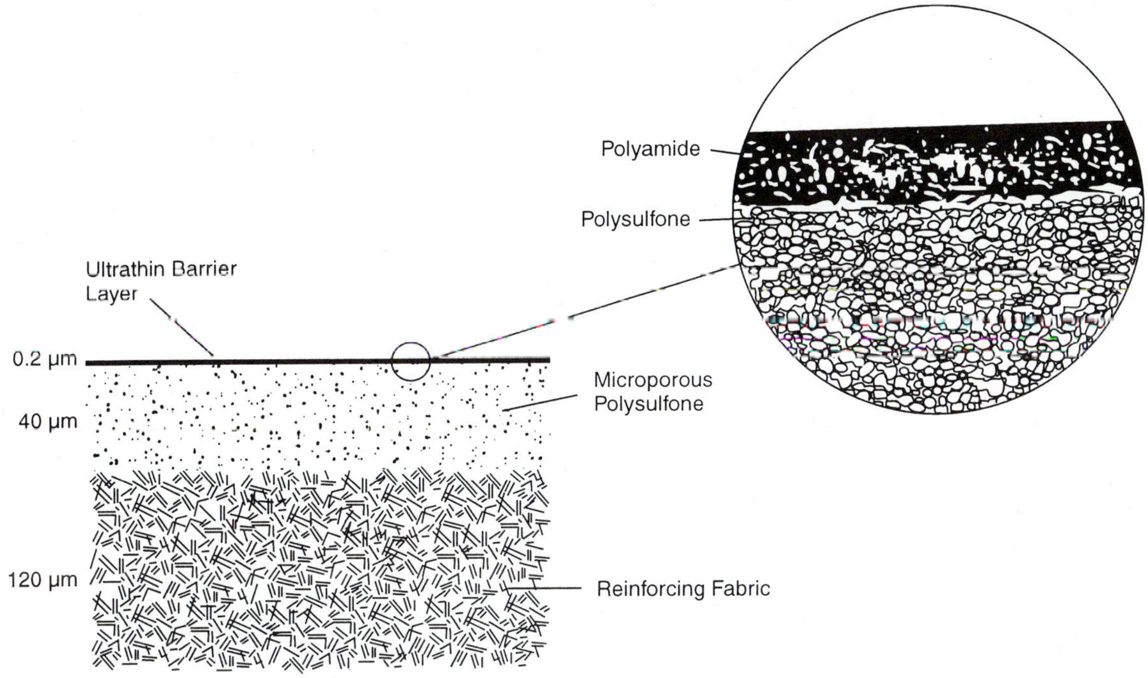

Figure 2-14 Schematic cross section of a thin-film composite reverse osmosis membrane

their reduced operating pressures, excellent divalent rejection, and partial SR, NF membranes are finding wide applications as the best available technology for removing trihalomethane (THM) precursors from hard, shallow groundwater and from certain surface water supplies.

A membrane by itself is quite fragile. It must be carefully supported and packaged so that it can be integrated into a unit process. In addition, to support and package the membrane, features are added that enhance permeation or water transport through the membrane (flux) and that lower the salt passage. The membrane is packaged to minimize hydraulic pressure losses and yet allow sufficient velocities to keep the surface of the membrane washed clean of concentrated salts. Mechanical design should also allow for passage of colloidal and particulate matter to minimize membrane fouling; such fouling will reduce flux and increase salt passage.

The following four basic package designs are in commercial use: tubular, plate and-frame, spiral-wound, and hollow fiber U-tube.

Tubular and Plate-and-Frame

Tubular and plate-and-frame membranes date back to the beginning of membrane technology. Both of these designs have a low membrane packing density, resulting in a higher capital cost than for spiral-wound or hollow fiber membranes. These designs are used in the food industry (e.g., whey concentration in cheese production and tomato juice concentration) and in wastewater treatment. They seldom compete with spiral-wound and hollow fiber devices in potable applications of brackish water or seawater desalination.

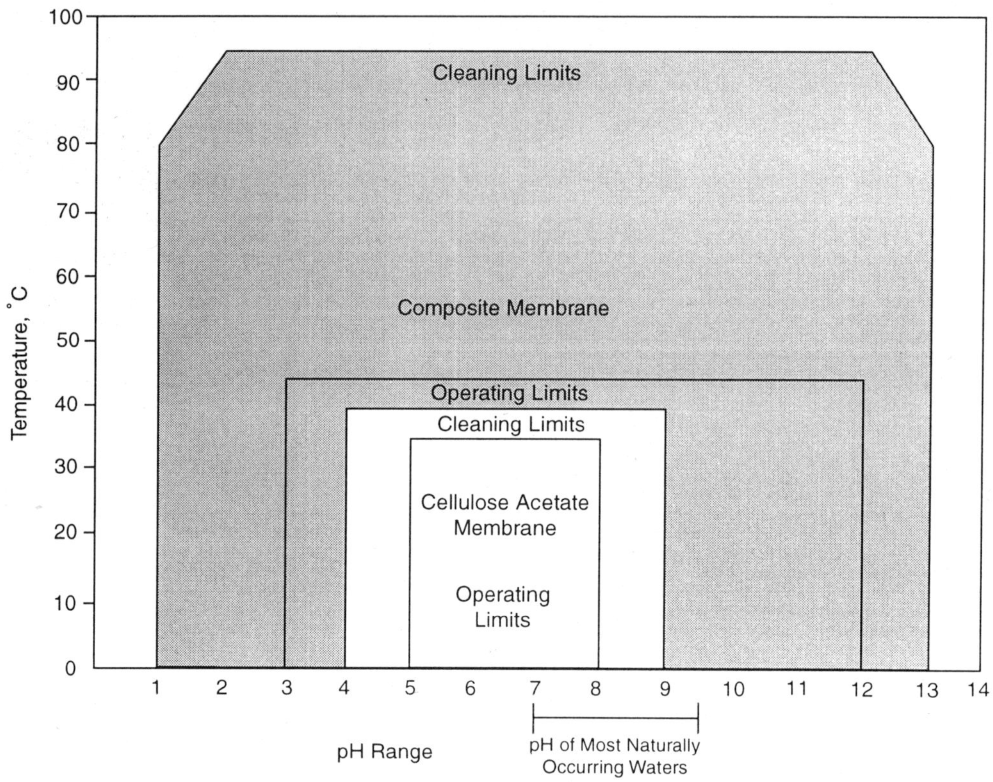

Figure 2-15 Comparison of the operating and cleaning parameters of composite membranes and cellulose acetate membranes

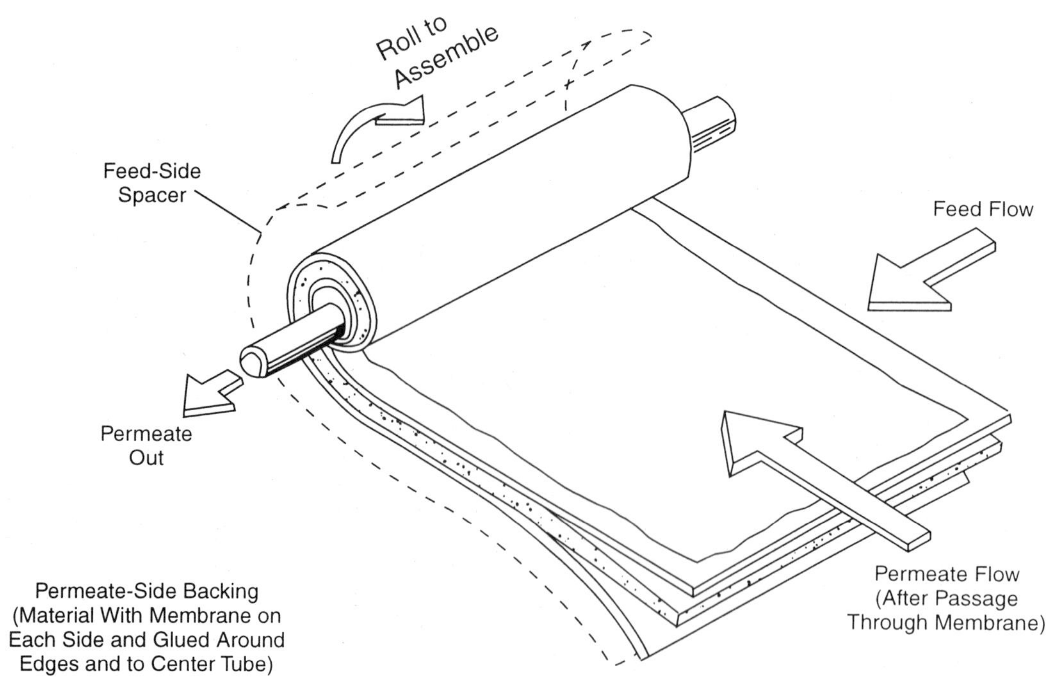

Figure 2-16 Spiral-wound module

Spiral Wound

A membrane for a spiral-wound device, shown in Figure 2-16, is manufactured or cast in sheet form on a backing material: sail cloth for cellulosic membranes and a nonwoven polyester web for the newer composite membranes. Two of these sheets are placed back to back, separated by a spacing fabric that acts as a permeate channel. Two sides and one end of this sandwich assembly are glued together to form an envelope or leaf. The open end of the leaf is connected to the permeate tube, around which the leaf is wrapped to form the spiral. An additional plastic netting is wrapped with the leaf to separate the membrane surfaces and maintain the feed-stream channel height. The spiral assembly, or element, is secured to prevent unraveling, and a brine concentrate seal is fixed to one end.

The elements are housed in a cylindrical vessel, illustrated in Figure 2-17, with the feed and concentrate flowing through the feed-side channels in a straight-axial path parallel to the direction of the permeate collection tube. Some of the water penetrates the membrane and "spirals" its way to the center, collecting in the permeate tube. The remaining water passes from the element and out the concentrate port of the pressure vessel. For most applications, several elements, usually six or seven, are housed in series within the pressure vessel. The concentrate from one element serves as the feed for the next in series.

Hollow Fiber

A hollow fiber membrane is spun with its own support structure. Hollow fibers as small as the size of a human hair (84 microns) are produced, with the wall thickness making up approximately one-half of the diameter. A diagram of a hollow fiber is shown in Figure 2-18. These fibers are bundled together as U tubes with the open ends potted in an epoxy tube sheet. The other end of the fiber bundle is also sealed in epoxy to prevent short-circuiting of the feed stream to the concentrate outlet. The bundle is cased in a pressure vessel with the pressurized feedwater distributed from a tube in the center of the bundle. As it flows radially through the bundle and over the fibers, some of the water penetrates the fibers, flows down the bore, and is collected at the tube sheet end of the vessel. The remainder of the water carries concentrated salts to the concentrate port of the vessel.

Because the membrane packaging density is so high, this device has the highest ratio of water production to space occupied. The low flux offers the advantage of minimizing the problem of concentration polarization. This high-density packaging

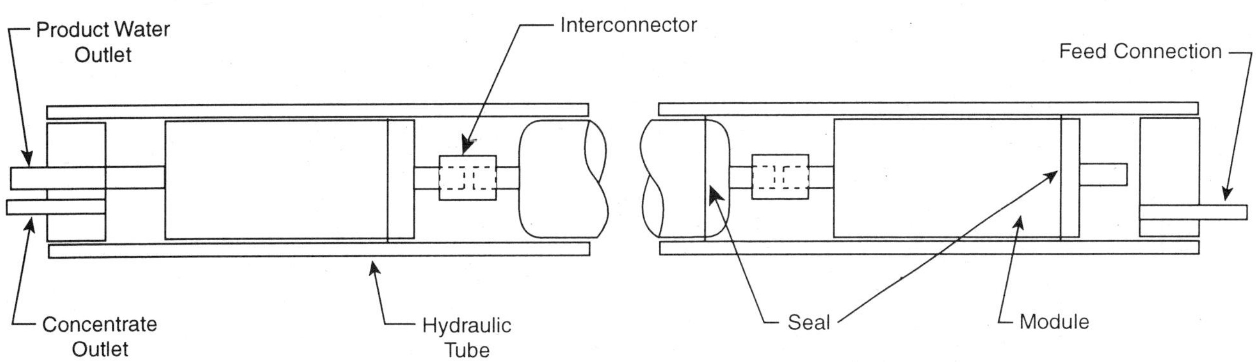

Figure 2-17 Pressure vessel assembly

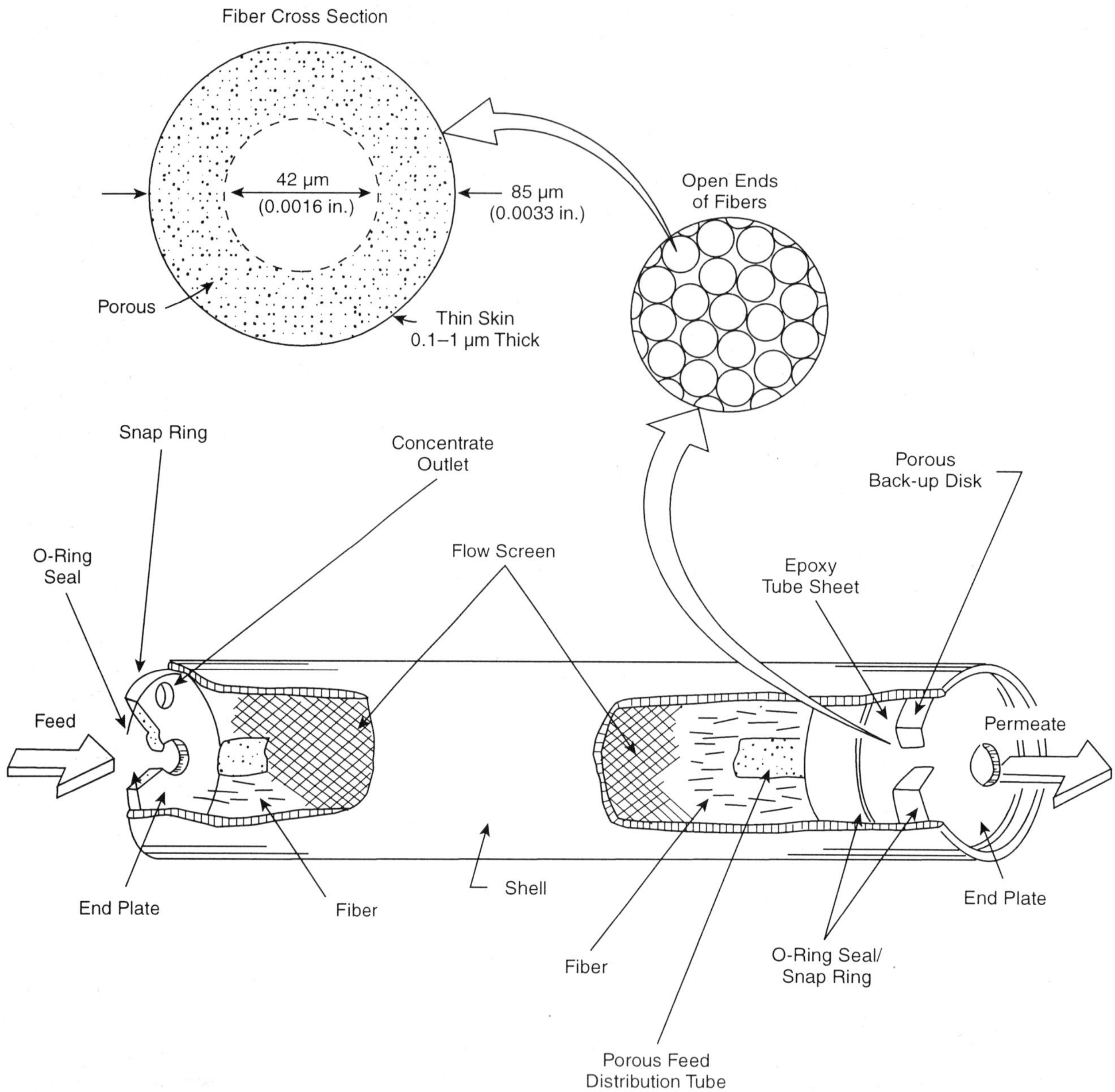

Figure 2-18 Hollow fiber U-tube permeator

leaves very little space between the fibers, making the size and amount of suspended solids and colloidal material more critical to satisfactory performance.

POSTTREATMENT

Reverse osmosis and NF processes produce corrosive finished waters because they reduce pH and typically remove too much calcium and alkalinity to provide a buffered and easily stabilized finished water. Consequently, posttreatment of the permeate water is required before distribution. (Microfiltration and UF processes do not reduce pH or remove any dissolved calcium or alkalinity. As a result, they do not directly affect the finished water buffering or stabilization.) A typical RO or NF membrane process for treating a groundwater supply is presented in Figure 1-3 on page 5.

Posttreatment for RO and NF processes in general can be extensive and depends to a large extent on the type and quality of the source water supply. The primary posttreatment operations covered in this section are alkalinity recovery, gas stripping, disinfection, and stabilization or corrosion control. Examples of typical water quality changes due to posttreatment are also illustrated for seawater, brackish water, a coastal Florida aquifer, and a coastal surficial Florida groundwater. In addition, appendix C lists the chemical reaction equations described in this section.

Alkalinity Recovery

Although actual corrosion rates are difficult to predict or model, maintaining a slightly positive LSI value has been shown to be beneficial in reducing general corrosion rates in most potable waters. Alkalinity and calcium hardness are two factors in an LSI calculation. The alkalinity concentration provides an indication of a water's resistance to pH changes on the addition of an acid. It is generally recommended that a finished water alkalinity greater than 40 mg/L as calcium carbonate be maintained within the distribution system to provide pH stability. Alkalinity is defined as the sum of the titratable bases and is dominated by the carbonate system in natural waters. Most alkalinity in natural waters is composed of hydroxide (OH^-), carbonates (CO_3^{2-}), and bicarbonates (HCO_3^-).

In a typical membrane process, the source water pH is decreased to between 5.5 and 7.0, depending on the membrane material, to prevent calcium carbonate from precipitating in the membrane. Figure 2-19 depicts the distribution of dissolved inorganic carbon forms, predominantly carbonic acid, bicarbonate ion, and carbonate ion, as a function of pH. Because the pH has been reduced, a substantial proportion of the alkalinity is converted to carbon dioxide and is lost to the atmosphere during the gas stripping phase of posttreatment.

Major reactions affecting alkalinity include dissolution of carbon dioxide gas and ionization of aqueous carbonic acid and bicarbonate. Because most membranes under normal operating conditions do not remove dissolved gases, substances such as carbon dioxide in anaerobic groundwater readily pass through the membrane. Reverse osmosis or NF systems using acid pretreatment will contain higher concentrations of carbon dioxide in the feed and the permeate because of the conversion of bicarbonate ion to carbon dioxide (carbonic acid when dissolved in water).

Acidifying during pretreatment is necessary so that the alkalinity in the feedwater can be converted to carbon dioxide, regardless of the required pH for scale control. If only antiscalant were used during pretreatment, this conversion would not occur. The carbon dioxide passing through the membrane then becomes the source of inorganic carbon for finished water alkalinity, so it is a vital part of producing a noncorrosive finished water.

Carbonates are found in many passivating scales, but calcium is not. Low total dissolved solids (TDS) levels in the permeate increase resistivity, but alkalinity helps passivate the corrosion reaction. If gas stripping follows directly after a membrane that received acid pretreatment, carbon dioxide will be lost to the atmosphere. Only a small percentage of the alkalinity that had been converted to carbon dioxide during pretreatment would remain. Subsequent stabilization with caustic would produce mostly hydroxide alkalinity and would result in a poorly buffered finished water.

Modifying the conventional membrane treatment system to include alkalinity recovery before gas stripping (not generally practiced) will produce a finished water with 1 to 3 meq of carbonate alkalinity in typical situations. Carbon dioxide passes directly through the membrane and can be converted to bicarbonate/carbonate alkalinity by raising the pH of the permeate before gas stripping. Sources of alkalinity include sodium hydroxide, calcium hydroxide, calcium oxide, or filtration through calcium carbonate. Using a base containing calcium will reduce the stabilization pH more than for other bases because of the accompanying addition of calcium during alkalinity recovery. If alkalinity recovery is not practiced, a highly corrosive water will be produced.

It should be noted that the LSI provides an indication of calcium carbonate precipitation potential in a particular water. Other corrosion indexes should be calculated and the specific water quality examined to fully evaluate the corrosivity of a finished water.

Gas Stripping

Degasification following membrane treatment is usually required to remove gases, such as carbon dioxide and hydrogen sulfide, that pass through NF and RO membranes and are present in the permeate. Several different types of aeration or gas transfer equipment are used for permeate degasification. The more common types of units are tray aerators, air-strippers, and packed towers.

A gas-stripping tower increases oxygen to the permeate, which increases the redox potential of the finished water. Since carbon dioxide is an acidic gas, its removal tends to increase the pH of the water passing through the tower. For example, a water with an alkalinity of 100 mg/L would have a pH of 7.6 after air-stripping. A water

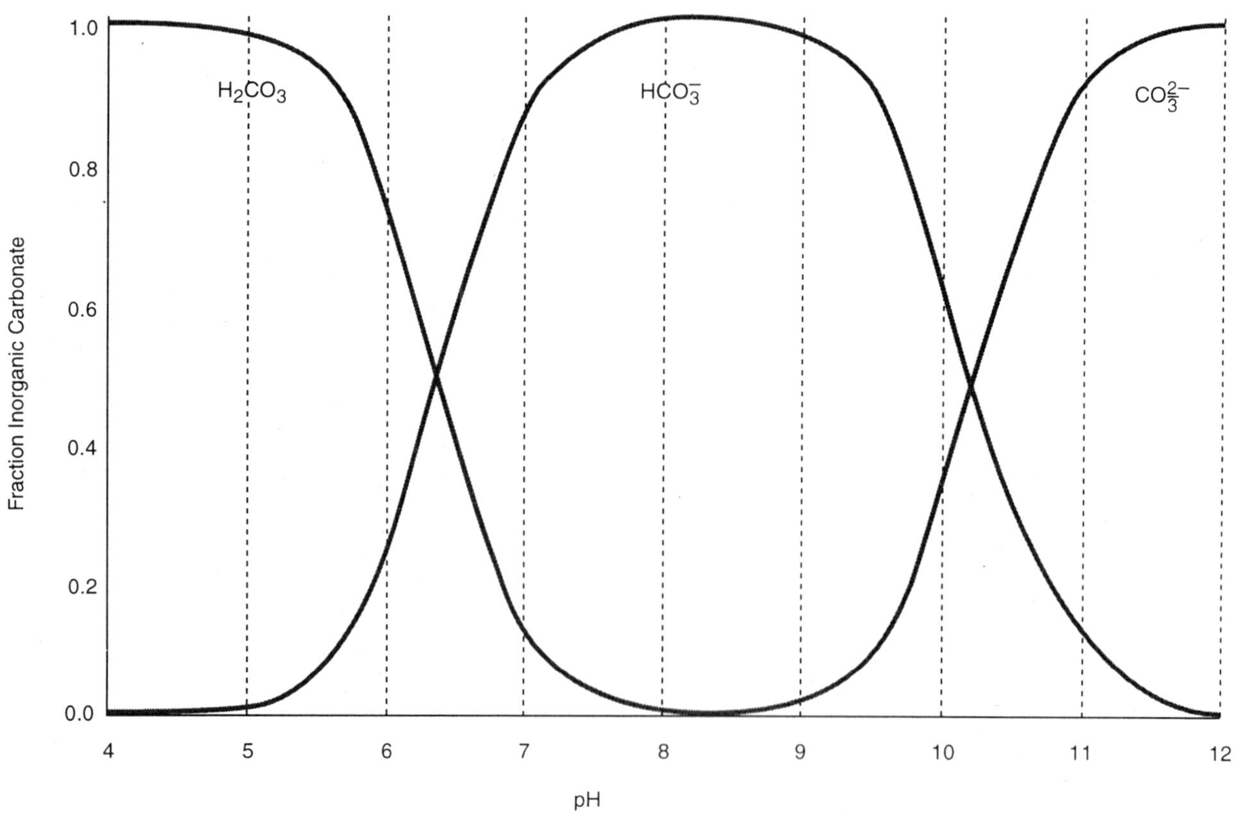

Figure 2-19 Inorganic carbonate distribution as a function of pH

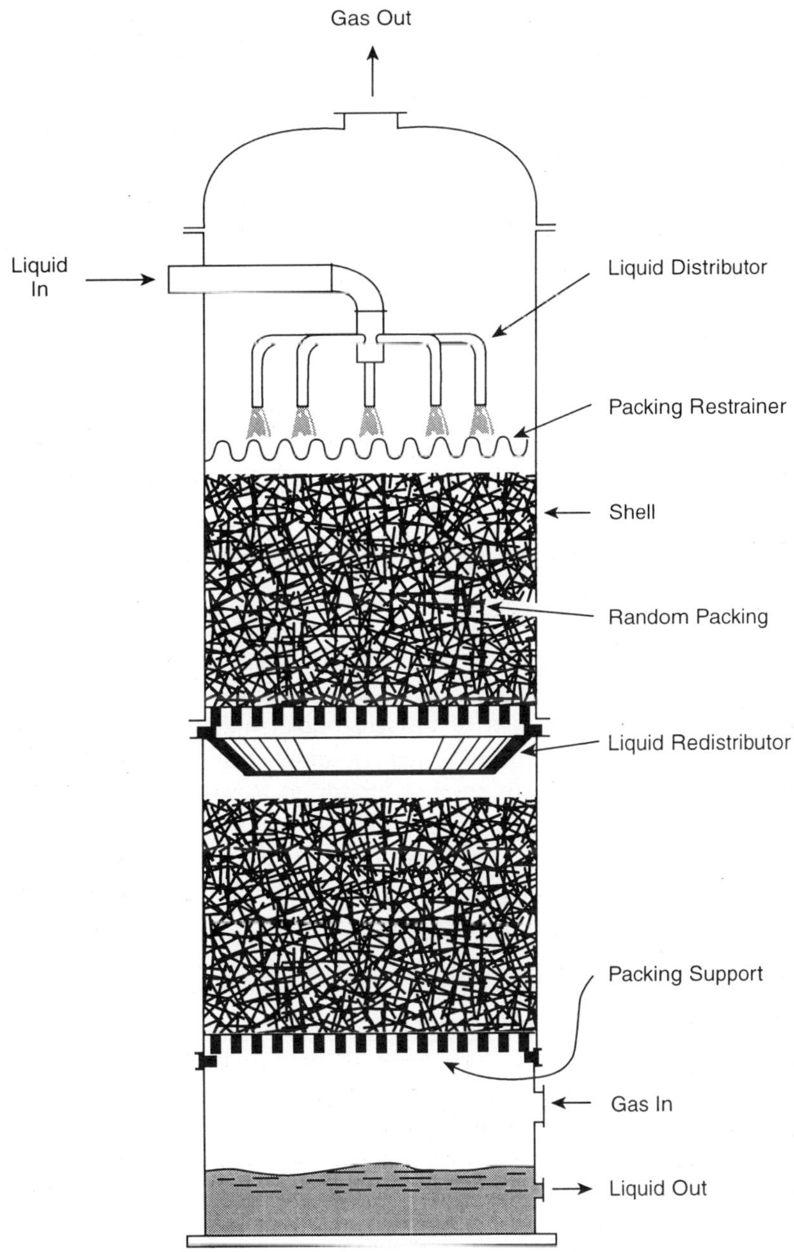

Figure 2-20 Typical gas-stripping packed tower

having a lower alkalinity would tend to have a lower pH on gas stripping, and a water having a higher alkalinity would tend to have a higher pH.

Cascading tray aerators consist of a series of trays with perforated bottoms. Water is distributed and falls into a collection basin at the base. Three to nine trays are typically used, with water application rates ranging from 20 to 30 gpm/ft^2. Tray aerators create additional chemical reactions that can have undesirable effects on the finished water quality. In the case of inefficient stripping of hydrogen sulfide, elemental sulfur turbidity can be produced.

Most modern RO and NF plants use packed towers for gas stripping. Packed towers consist of a cylindrical shell containing a plate that supports the packing material, with water and gas flowing in a countercurrent pattern, as shown in

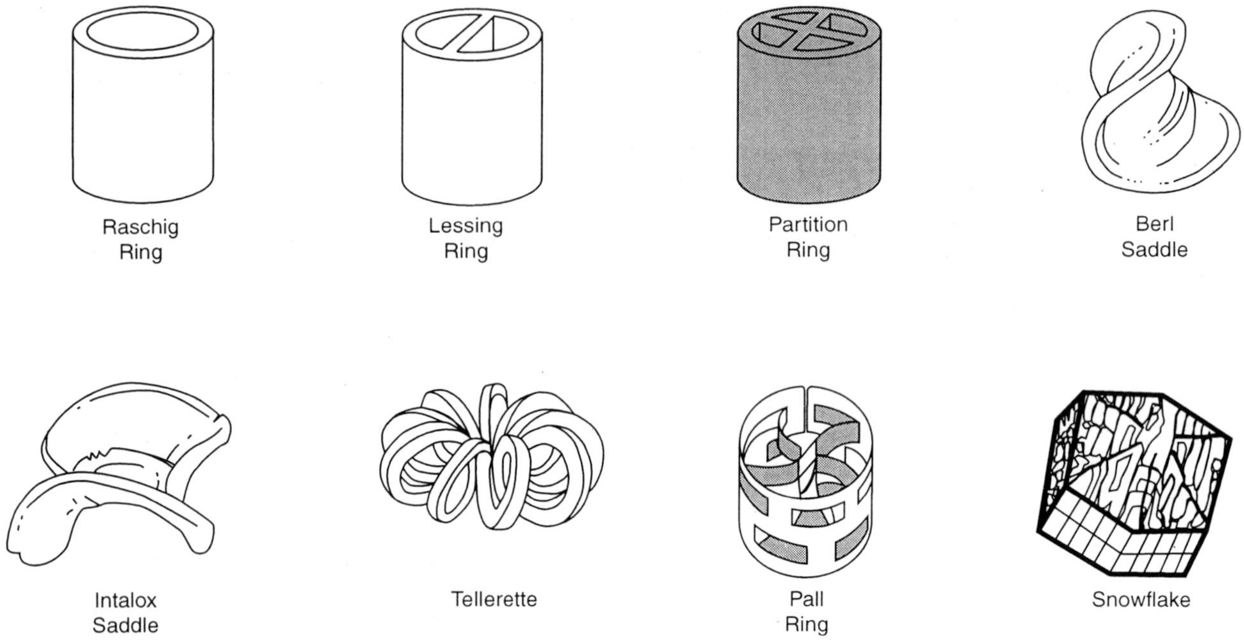

Figure 2-21 Various packing materials

Figure 2-20. Various types of plastic materials used for packing are illustrated in Figure 2-21. Although not typically used because of the low permeate scaling potential, antiscalant may need to be added before the packed tower to avoid calcium carbonate scaling of the packing.

For source waters that contain hydrogen sulfide, gas stripping removes objectionable tastes and odors and prevents colloidal sulfur turbidity. If a plant is near residences, an off-gas scrubber or odor control system should be provided. The amount of total sulfide dissolved in the water in ionic form is directly dependent on pH. Figure 2-22 shows that sulfide can exist as a dissolved gas (H_2S), as a protonated ionic form (HS^-), or as sulfide ion (S^{2-}). For total sulfide to be effectively removed by gas stripping, it must be in the gaseous form as hydrogen sulfide, H_2S. At a pH of 8.0, for example, only 10 percent of the total sulfide present is gaseous, and at pH 7.0 about 50 percent exists as hydrogen sulfide. Tray aerators remove between 40 and 70 percent of sulfides, so they are comparatively inefficient relative to packed towers, which achieve greater than 90 percent removal efficiencies of hydrogen sulfide for an influent pH of approximately 6.

More important, the pH before the use of aeration to remove hydrogen sulfide should not be less than the pH that retains the desired alkalinity, which is a function of the carbon dioxide that passes through the membrane. For example, adjusting the pH to 6.3 before gas stripping will provide approximately 1 meq, or 50 mg/L, of alkalinity as calcium carbonate. Consequently, finished water goals should take into account the optimal hydrogen sulfide removal without sacrificing alkalinity recovery and should be determined on a case-by-case basis.

The main problems with air-stripper scrubber systems are the following:

- Increasing the dissolved oxygen concentration in finished water may increase corrosion.
- Venting hydrogen sulfide to the atmosphere may cause objectionable odors and increase structural corrosion of buildings and equipment.

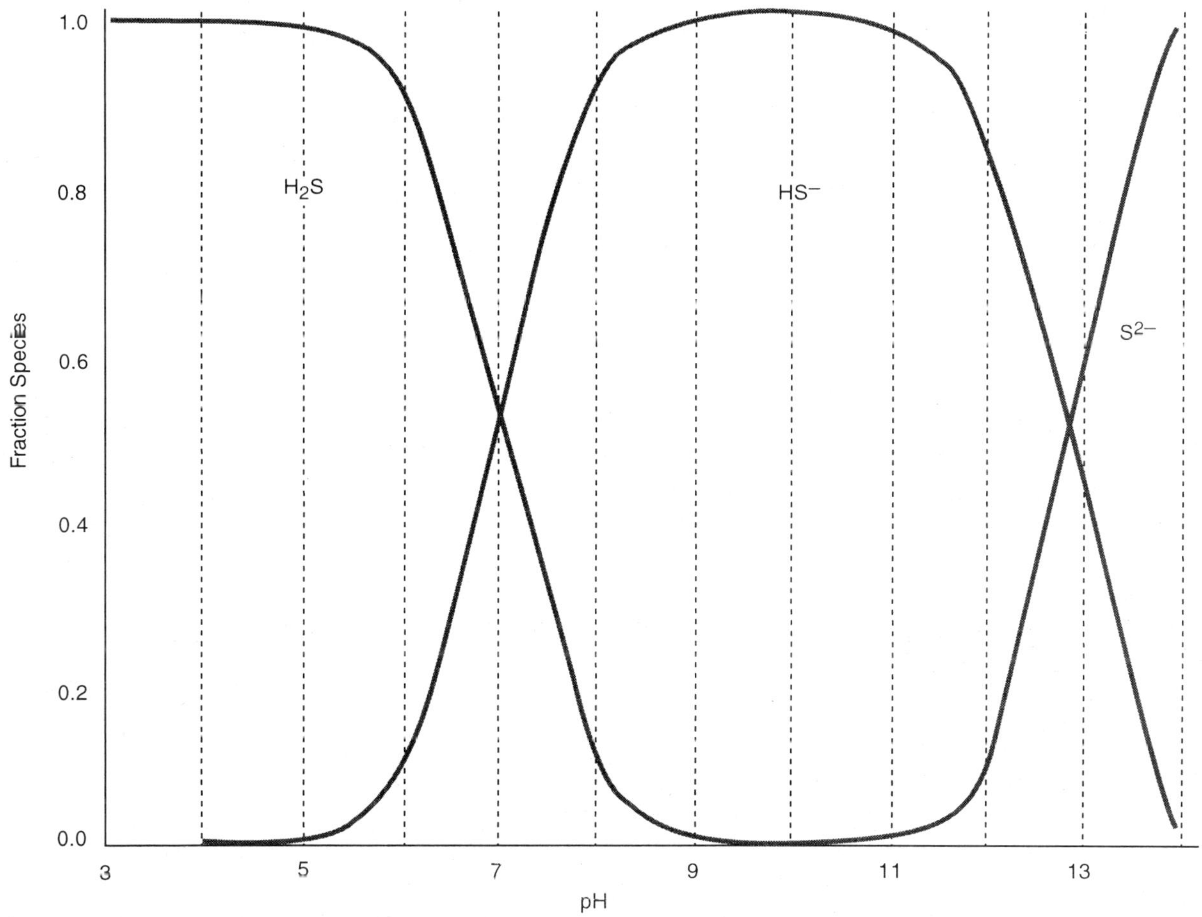

Figure 2-22 Sulfide species distribution for varying pH

- Disposing of the scrubbate waste from an odor-control system may create additional problems.

Hydrogen sulfide can be converted directly to sulfates in the permeate water without the need for air-stripping. Reverse osmosis permeate water can be treated with chlorine at a pH of approximately 3.0 to convert the hydrogen sulfide completely to sulfate without colloidal elemental sulfur turbidity being produced. In addition, NF permeate water has been treated with chlorine at pH 3.8 with similar results. In these cases, care must be taken to avoid producing obnoxious by-products such as nitrogen trichloride, NCl_3, because the nature of these side reactions is dominated by free chlorine. The finished water is then stabilized with the addition of sodium hydroxide or another appropriate chemical. Because the permeate is not aerated, dissolved oxygen levels remain less than 2 mg/L, as would be expected of an anaerobic groundwater source. Off-gas treatment is not required because all the hydrogen sulfide has been converted to sulfate. (Large amounts of chlorine can be required in an off-gas process, approximately 8.4 lb chlorine per pound of hydrogen sulfide.) Since the membrane has already rejected the majority of the disinfection by-product precursors, DBPs and THMs are minimal. In addition, all carbonate alkalinity is retained because carbon dioxide remains within this closed system. However, the additional chemical costs for acidification, chlorine oxidation, and subsequent caustic addition must be compared against costs for packed tower stripping.

Disinfection

Disinfection is required to destroy or reduce pathogenic microorganisms before distribution. The practice of chlorinating US public water supplies for disinfection dates back to 1904, although chlorine has been used on an emergency basis since the 1850s. Although the mechanism of disinfection is not well understood, the active disinfectant is hypochlorous acid regardless of whether chlorine gas or sodium hypochlorite is initially used.

When chlorine gas dissolves in water, it instantaneously reacts to form hydrochloric acid (which completely dissociates) and hypochlorous acid. Hypochlorous acid is a weak acid. The extent of dissociation is thus dependent on pH, temperature, and concentration. The sum of the hypochlorous acid and hypochlorite ion in solution is the *free available chlorine*. Maintaining a free chlorine residual is the principal method water utilities use to ensure proper disinfection. Adding chlorine to water increases chlorides and, hence, TDS in the finished water. In addition, in waters where hydrogen sulfide is present, chlorine will react with it to form elemental sulfur, hydrochloric acid, and sulfuric acid. Elemental sulfur can be responsible for increased turbidity. The chlorine requirement for complete oxidation of hydrogen sulfide is four times that for sulfur precipitation.

The advantage of using chlorine to disinfect RO and NF permeate water is that membranes reject DBP precursors and other oxidizable substances from the feedwater. As a result, the use of RO or NF decreases the chlorine demand in the permeate water, thereby reducing the cost of disinfection. Typical dosages of 1 to 4 mg/L are required for permeate disinfection, depending on distribution system constraints and water quality.

In conventional (nonmembrane) treatment, ozone, chlorine dioxide, and chloramines are viable alternatives to chlorine for disinfection. These alternatives are typically used to avoid the formation of DBPs. Because membrane treatment almost completely removes DBP precursors, chlorine is the most logical and widely used chemical for permeate disinfection. However, distribution system conditions should be considered in determining the most appropriate system residual disinfectant, and the other disinfectants could be suitable.

Stabilization and Corrosion Control

Stabilization produces a permeate water that tends to be noncorrosive and typically has a pH between 6.5 and 8.5. Stabilization is usually required for RO permeates because of reduced pH, bicarbonate alkalinity, and calcium concentrations. For NF systems using groundwaters high in sulfide, stabilization is also required if air-stripping is used to remove hydrogen sulfide.

A finished water is considered to be stable when it is neither undersaturated nor supersaturated with respect to calcium carbonate. The process of determining the point at which saturation exists is based on the work of Langelier in 1936 and on identifying the pH of saturation, or pH_S. An equation to calculate pH_S is found in appendix E, and graphs and computer software are available to determine its value.

Several indexes, the most common being the LSI, are used to determine if a permeate water is non–scale-forming, neutral, or scale-forming with respect to calcium carbonate. Positive LSI values generally indicate a scale-forming water, and negative values represent a non–scale-forming water. However, this statement is a generalization, and corrosion potential is a complex topic subject to specific water quality. Other factors, such as dissolved oxygen level, chloride concentrations, and temperature, play critical roles in determining the corrosivity of a particular water. It should be emphasized that calcium carbonate scaling potential is only a general indicator of corrosion potential. Corrosion potential should be considered on a site-specific basis, with a full evaluation of all water quality parameters that could contribute to corrosion of metallic pipe materials.

The LSI may be calculated as follows:

$$LSI = pH - pH_S \tag{2-3}$$

Where:

LSI	=	Langelier Saturation index, dimensionless
pH	=	pH of the water being considered
pH_S	=	pH of saturation

The aggressivity index (AI) was developed at the request of consulting engineers to govern the selection of the proper type of asbestos–cement pipe and to ensure long-term structural integrity. This index should not be applied to other pipe materials. The AI is defined by AWWA C-400, Standard for Asbestos–Cement Pressure Pipe, as follows:

$$AI = pH + \log(AH) \tag{2-4}$$

Where:

AI	=	aggressivity index, dimensionless
A	=	total alkalinity, in mg/L as $CaCO_3$
H	=	calcium hardness, in mg/L as $CaCO_3$

The values obtained are interpreted as follows: $AI < 10$, very aggressive (corrosive); $10 \leq AI < 12$, moderately aggressive; $AI > 12$, nonaggressive. The AI is based on pH and the solubility of calcium carbonate. It is a simplified form of the LSI and approximates only the solubility of calcium carbonate. However, it can be a useful tool in selecting materials or treatment options for corrosion control for asbestos–cement pipe.

Methods typically used for stabilizing a corrosive permeate water include

- Degasification to remove carbon dioxide and hydrogen sulfide and to raise the pH of the water.
- Blending permeate water with bypass feedwater to increase the calcium content and alkalinity of the water.
- Adding chemicals to increase alkalinity and pH, or adding corrosion inhibitor chemicals. Permeate water can be stabilized by adding lime, calcium carbonate, caustic soda, or soda ash. Lime is the least expensive chemical, but its application represents increased capital costs for feed systems relative to other chemicals.

Filters packed with alkaline media, such as dolomite and calcite, can be used for anticorrosion treatment for small systems. An example might be alkaline filter media consisting of 66 percent calcium carbonate, 30 percent magnesium oxide, 2.5 percent magnesium carbonate, and 1.5 percent silicium dioxide. The porous, crystal-hard,

Table 2-20 Typical water quality for a seawater reverse osmosis membrane process

Parameter	Source, *mg/L*	Permeate,* *mg/L*	Finished, *mg/L*
Calcium	400	1.4	1.4
Magnesium	1,292	4.6	4.6
Sodium	10,734	105	109.7
Potassium	385	3.4	3.4
Strontium	14	0	0
Bicarbonate	144	4	16.6
Sulfate	2,688	9.6	9.6
Chloride	19,336	171.6	171.6
Fluoride	1	0	0
Carbon dioxide	2.4	8.9	0.14
TDS	35,000	299.6	316.9
pH	8†	5.9†	8.2†

*Results obtained for 50 percent recovery, 1,000 psig, 20°C.

†Dimensionless.

round grains of the material provide good filter media, with which water reacts both chemically to increase alkalinity and pH and physically to remove turbidity. These media add only calcium, magnesium, and bicarbonate ions to water during treatment to provide an equilibrated, noncorrosive finished water.

For membrane processes that first use packed tower air-strippers to remove carbon dioxide and hydrogen sulfide and then use caustic or soda ash to increase the pH to saturation, only hydroxide alkalinity will remain in the finished water. Although the resultant pH value may produce a positive LSI, this does not mean that the permeate water will be noncorrosive. Actually, all calcium has been removed. Thus, if caustic or soda ash are added for pH adjustment, then inhibitors such as phosphates should be used for corrosion control.

Reverse Osmosis and Nanofiltration Posttreated Water Quality

After posttreatment, a fully stabilized and disinfected water must be achieved. This water should meet all applicable potable water regulations. Local water usage patterns can also have an impact on the desired finished water quality. For example, there might be a local preference concerning the degree of total hardness in the finished water.

The following examples illustrate the results of posttreatment on finished water quality.

Example 1: Seawater. Table 2-20 lists typical results for a seawater RO process. The carbon dioxide content of seawater at pH values normally seen in surface intakes is usually low. Hydrogen sulfide is normally not present in seawater, although it can be found off the Florida coast. In feedwaters lacking these gases, gas stripping is not required. Bicarbonate alkalinity and hardness are usually low and may have to be increased for corrosion control by the addition of bicarbonate, compounds containing carbonate, and/or hardness, usually as calcium.

Example 2: Brackish water. Brackish water RO has frequently been used in southwest Florida. Table 2-21 indicates water quality data for such a facility. Posttreatment chemicals after degasification for this example consist of caustic, zinc orthophosphate, ammonia, and chlorine.

Example 3: Coastal surficial groundwater. Table 2-22 lists water quality data for a coastal Florida aquifer RO plant. Pretreatment consists of adding a scale inhibitor and sulfuric acid to lower the pH to approximately 6.0. Because of the

Table 2-21 Typical water quality for a brackish water reverse osmosis membrane process

Parameter	Source, *mg/L*	Permeate,* *mg/L*	Finished, *mg/L*
Calcium	131	2	2
Sodium	1,725	125	129
Chloride	2,978	193	193
Bicarbonate	197	14.3	20.1
Carbonate	0.2	0	0
TDS	5,736	351	360
pH	7.4†	5.6†	8.3†

*Posttreatment includes degasification, as well as caustic, orthophosphate, and chlorine addition.
†Dimensionless.

Table 2-22 Typical water quality for a brackish water aquifer reverse osmosis plant in coastal Florida

Parameter	Source, *mg/L*	Permeate, *mg/L*	Finished,* *mg/L*
Calcium	102	3.5	32
Sodium	235	39	42
Chloride	400	44	50
Bicarbonate	215	7	8.3
Carbonate	0	0	5.2
Carbon dioxide	6	57	0
TDS	983	65	255
pH	7.8†	5.5†	8.3†

*Permeate is mixed with lime-softened water after gas stripping.
†Dimensionless.

Table 2-23 Water quality for a membrane softening facility treating coastal surficial Florida groundwater

Parameter	Source, *mg/L*	Permeate,* *mg/L*	Finished, *mg/L*
Calcium	147	17	17
Magnesium	11	1	1
Sodium	25	18	18
Chloride	66	27	27
Bicarbonate	372	39	41
Carbonate	0	0	0.5
Carbon dioxide	25	68	0
TDS	390	85	91
THM formation potential	537	15	19
pH	7.1†	6.5†	7.8†
Color	58	2	1

*Typical NF membrane, first and second stages; typical RO membrane, third stage.
†Dimensionless.

presence of carbon dioxide and hydrogen sulfide in the permeate, gas stripping is used. There is a loss of alkalinity because the preacidification step converts all bicarbonate to carbon dioxide gas, which on aeration is lost to the atmosphere. Posttreatment includes blending permeate with lime-softened water before disinfection.

Example 4: Coastal surficial groundwater. Table 2-23 presents the results of a membrane softening facility in southwest Florida. This water has high color, total organic carbon, and hydrogen sulfide, with a THM formation potential of over 500 mg/L. As in the case of example 2, gas stripping without alkalinity recovery is practiced, so the bicarbonate buffer is lost.

CONCENTRATE DISPOSAL

Most US membrane facilities do not reduce the amount of concentrate stream. Methods used to do this include solar ponds, thermal evaporators, crystallization techniques, solar distillation, and electrodialysis. Although all these techniques have been used successfully, most are too costly or too site-specific for widespread use. In addition, all the methods still require disposal of a highly concentrated brine or salt solids.

The membrane plant survey discussed in chapter 1 gathered information concerning concentrate disposal techniques. Table 2-24 presents the recoveries, concentrate flows, disposal methods, and flows for the 20 RO or NF membrane plants that responded to the plant survey. Operating pressures for plants surveyed ranged from 125 to 400 psi, with the average being 360 psi. Based on the survey, certain concentrate disposal methods, i.e., saline water discharge, deep well injection, irrigation, and percolation via infiltration basins, were selected for discussion here. For each method, a brief history, the major regulatory requirements, a technical description, and the costs will be presented in the following text. The reader should also consult Mickley et al. (1993) for further information.

Table 2-24 Recoveries, concentrate flows, and disposal methods for plants responding to survey

Plant Number	Recovery, %	Flow, *1,000 gpd*	Disposal Method
1	50	40	Ocean outfall
2	82	1,460	Tidal lake
3	75	91	Saltwater canal
4	50	–	Saltwater canal
5	40	1.3	Infiltration basin
6	85	245	Brackish water canal
7	70	130	Spray irrigation
8	80	605	Ocean outfall
9	75	283	Sewer discharge
10	40	0.8	Sewer discharge
11	70	100	Ocean outfall
12	75	60	Land spreading
13	–	–	–
14	45	–	Deep well injection
15	33	–	Spray irrigation
16	50	500	Ocean outfall
17	75	1,500	Ocean outfall
18	50	–	Land spreading
19	55	1,000	Saltwater spray
20	42	2	Land spreading

Source: Taylor et al. (1989).

Saline Water Discharge

Saline water discharge involves transporting concentrate from the membrane plant to a point of outfall. The point of outfall may be a bay, tidal lake, brackish canal, or ocean.

History and regulations. At the first membrane facilities in Florida, concentrate disposal was relatively simple—all plants were near the coast, and concentrate was discharged directly into the sea or a brackish canal. However, increased regulatory pressure, public concern over any surface water discharges, and the development of inland membrane facilities may cause surface water discharge to become a less attractive and less economically feasible concentrate disposal option.

The location of the point of outfall and the need for concentrate treatment are determined by state and USEPA maximum contaminant levels for surface water discharge. Concentrate disposal is usually affected by all regulations written to protect groundwater and surface water.

To determine if concentrate treatment will be necessary, pilot studies can be conducted or a method of theoretically determining concentrate quality can be used. Actual concentrate water will be required in order to conduct bioassay tests. Toxicity concerns often determine whether specific concentrate disposal methods may be permitted. The following rules theoretically predict concentrate quality based on source water characteristics for the purposes of preliminary or feasibility studies:

- Heavy metals will be rejected in a similar ratio as calcium and magnesium.
- Organics are typically rejected in excess of 95 percent (except for low-molecular-weight organics).
- Concentrate will be anaerobic and contain hydrogen sulfide. (Nearly all the groundwater used at membrane plants in Florida is anaerobic and contains hydrogen sulfide.)
- Concentrate pH is generally higher than the source water pH because of the greater concentration of alkalinity in the concentrate.
- For brackish water RO, the ion concentration factor based on 100 percent (worst case) SR can be calculated from the following equation:

$$CF = \frac{1}{1 - Y} \tag{2-5}$$

Where:

CF = concentration factor, dimensionless
Y = recovery, expressed as a decimal

Concentrate quality can also be predicted based on a diffusion model if the operating pressure, source water quality, percent recovery, and solvent and solute mass transfer coefficients are known. The equations used are found in the theory section of chapter 1 and in appendix C.

Technical description and cost. The concentrate may be transported via an open canal or an enclosed pipeline system. Pumping may or may not be required depending on the concentrate head exiting the membrane and frictional forces encountered during transport. The outfall mechanism usually consists of an open-ended pipe or a perforated pipe, as well as a surge control system.

The cost of concentrate disposal through a saline water discharge system is affected by the concentrate flow, regulatory concerns regarding treatment, method of transport, operating costs, and outfall mechanisms.

Table 2-25 Pretreatment cost analysis for saline water discharge*

	Capacity and Cost			
Component	1.0 mgd, *\$1,000*	3.0 mgd, *\$1,000*	6.0 mgd, *\$1,000*	12.0 mgd, *\$1,000*
Concentrate pipeline	105	174.4	244.1	348.8
Aerator	5.9	9.8	14.6	21.6
Transfer pump	14.4	24.4	29.7	34.3
Mixer	2.7	4.5	6.6	9.9
Total cost	128.0	213.1	295.0	414.6
Percent of total plant cost	10.0	9.0	7.0	5.5

*This table assumes the following conditions:
- discharge point 1.5 miles from the plant
- RO plant recovery at 85 percent
- PVC pipeline
- installed costs
- mid-1988 cost basis

Source: Morin and Malaxos (1988).

Concentrate discharged into a saline surface water body must meet state water quality standards; as a result, it may be necessary to treat the concentrate before disposal. Such treatment could include aeration, disinfection, and subsequent dechlorination. A cost analysis of concentrate treatment in an RO plant was performed in the late 1980s. The treatment consisted of aeration, transfer, and mixing. The results of this cost study are presented in Table 2-25. As the table indicates, concentrate treatment adds to the total membrane plant cost, although pipeline costs usually represent only a small capital cost factor. Factors affecting transport costs include the distance from the plant to the disposal site, the concentrate volume, and pipeline, pumping, and other operating costs. All historical cost analyses should be adjusted for inflation to provide current cost estimates.

Mixing the concentrate stream with the receiving water body at the point of discharge (i.e., the mixing zone) dilutes the concentrate. If this method is acceptable for regulatory compliance, the dimensions of the mixing zone must be approved during the permitting process. Regulators often establish the distinction between acute and chronic toxicity at the point of discharge.

Transport pipeline. The concentrate disposal site should be located as near to the membrane facility as possible. Transport pipe is priced on a per-foot basis, and pipeline capital costs increase with each increase in total pipeline length. The volume of concentrate is also an important cost factor because larger flows require larger-diameter pipes and pipeline prices increase as a function of pipe diameter.

Concentrate discharge pipes should be made of corrosion- and crush-resistant pipe. Cast iron has historically been one of the most popular of all pipe materials; it has corrosion-resistant properties and strength that approaches steel pipe. For concentrate streams containing a high chloride concentration, however, the corrosion of cast iron could be accelerated.

Steel pipe is lighter, normally less expensive, and easier to construct than cast-iron pipe. Another major advantage of steel is its ability to withstand higher pressures. Unprotected steel is highly susceptible to oxidation and galvanic corrosion. Pipe coatings are used on steel pipes to solve oxidation problems and cathodic protection is used to reduce galvanic corrosion.

Table 2-26 Plastic piping materials

Plastic Material	Acceptable Maximum Diameter, *in.*
Fiberglass-reinforced epoxy	144
High-density polyethylene	48
Polybutylene	24
Polyvinyl chloride	24

Developments in plastics technology have made possible the manufacture of larger-diameter plastic pipe. Table 2-26 lists a few examples of plastic pipe materials and their accepted maximum diameters. Polyvinyl chloride is a polymerized vinyl chloride derived from acetylene and anhydrous chloride. Polyvinyl chloride piping has replaced traditional materials of construction in many piping systems. Its advantages include its corrosion resistance, chemical inertness, light weight, and resistance to galvanic attack. In certain sizes, plastic piping is also less costly than other piping materials, such as cast iron.

In addition to the pipe material selected for pipeline construction, the choice of installing pipelines above or below ground impacts the overall cost of the transport system. Belowground piping requires excavation, whereas aboveground piping systems require support structures to be constructed in carefully compacted fill. Unusual ground conditions can significantly increase the cost of pipeline system installation.

Pumping. The concentrate flow resulting from a typical membrane system exits the system at pressures ranging from 50 to 200 psi or more. Under certain conditions, the resulting concentrate head is sufficient to overcome frictional losses within the pipe, allowing transport of the concentrate flow to the disposal site without the need for additional pumping. For example, in Plantation, Fla., a recently designed deep well injection system necessitated the concentrate to be transported 1,000 ft to the well. Instead of providing pumping, the design engineers used the pressure head that existed as the concentrate left the membrane system for transportation to and down the injection well. In other cases, though, the pressure head resulting from the membrane system is insufficient to overcome frictional losses, and additional pumping is necessary to transport the concentrate to the disposal site.

The need for additional pumping depends on the concentrate head exiting the membrane, the pipeline distance, and the transport head losses resulting from frictional forces. When pumping is necessary, the energy and maintenance costs associated with operating transport pumps become important cost factors. Surge control must also be considered in the design.

Outfall mechanism. Outfall mechanisms consist of a multiport diffuser or a simple open end. A multiport diffuser is designed so that the end of the transport pipe is capped and the last sections of the pipe contain lines of small ports (openings or diffuser nozzles around the circumference of the pipe). The purpose of the diffuser is to provide a greater initial dilution of the concentrate as it enters the saline surface water.

Most small outfalls, as well as larger ones built before 1980, have simple open ends. In recent years, multiport diffusers have become the accepted design for larger-diameter outfalls. Simple open-end outfalls are recommended when the initial dilution that is achieved meets water quality standards. If standards are not met, the more expensive—and more difficult to build—diffusers become necessary.

Deep Well Injection

In deep well injection, the waste stream is injected hundreds to thousands of feet below the ground surface into an injection zone. Injection zones contain natural fluids with high levels of TDS. Injection zones are overlayed with thick, impermeable layers of rock and other materials that prevent the injected concentrate from migrating upward toward the surface. The concentrate waste stream is permanently stored in the injection zone.

Membrane concentrate is classified by USEPA as an industrial waste and, as such, requires additional precautions for deep well construction. More specifically, deep wells disposing of industrial wastes are required to have an interior tubing liner and surrounding packer. For facilities in Florida, the Florida Department of Environmental Protection also limits the velocity of flow through the well to a maximum of 8 ft/s. This maximum velocity regulation, along with the concentrate flow rate, defines the minimum acceptable diameter of the well. Injection zones must have a TDS level greater than 10,000 mg/L and at least one overlaying, confining layer.

History and regulations. Deep well injection systems were developed by the petroleum industry in the early part of this century to dispose of oil field brines. These brines were injected into subsurface saline aquifers. Today, municipalities and industries dispose of large volumes of waterborne wastes in injection wells. Multiple layers of corrosion-resistant metal casing and cement are used to seal the well shafts.

Although wells are designed to prevent leakage, wells across the United States reportedly failed and contaminated groundwater. In response to these reports, the Underground Injection Practices Council (UIPC) in the 1980s retained a consulting engineering firm to investigate reported class I injection well malfunctions at 45 sites containing 106 wells. Class I injection wells are designed to dispose of industrial and municipal wastes. The final report indicated that most of the problems found were related to outmoded designs and practices not allowed under current minimum federal regulations. The UIPC concluded from the results of its extensive investigation that deep well injection of waterborne waste streams was environmentally sound when properly conducted and adequately regulated.

Class I and class V injection wells have been used in Florida since 1963, and 128 deep injection wells are proposed or currently operating in that state. Although most of these wells are used to dispose of treated municipal wastewater effluents, a few wells dispose of industrial wastes and membrane concentrate from RO and NF facilities. The ability of injection wells to operate under varying weather conditions is a major advantage.

Technical description. A deep injection well consists primarily of a pump (if needed) and a lined well shaft protected by multiple layers of casing and cement grouting. The overall cost of a deep well injection system is affected by the following main factors: concentrate volume, distance from the treatment plant to the well, transport pipeline, well depth, pumping pressure, storage, regulatory monitoring, and operation. Distance from the treatment plant and the factors affecting the transport pipeline are the same for a deep well system as for a saline water discharge system.

Injection well shaft. The type of fluid waste being injected should be compatible with the materials selected for well shaft construction. Concentrate from RO plants requires additional corrosion protection because of its generally high TDS levels. Materials often used for the inner liner of a well shaft include fiberglass, plastic, stainless steel, and extra-thick steel pipe.

Injection wells are generally constructed by the same process used to construct extraction wells. Cable-tool and rotary drilling have been used successfully to construct deep wells. Completion of the well involves testing the casing and cement grouting to make sure they do not leak and can sustain design pressures.

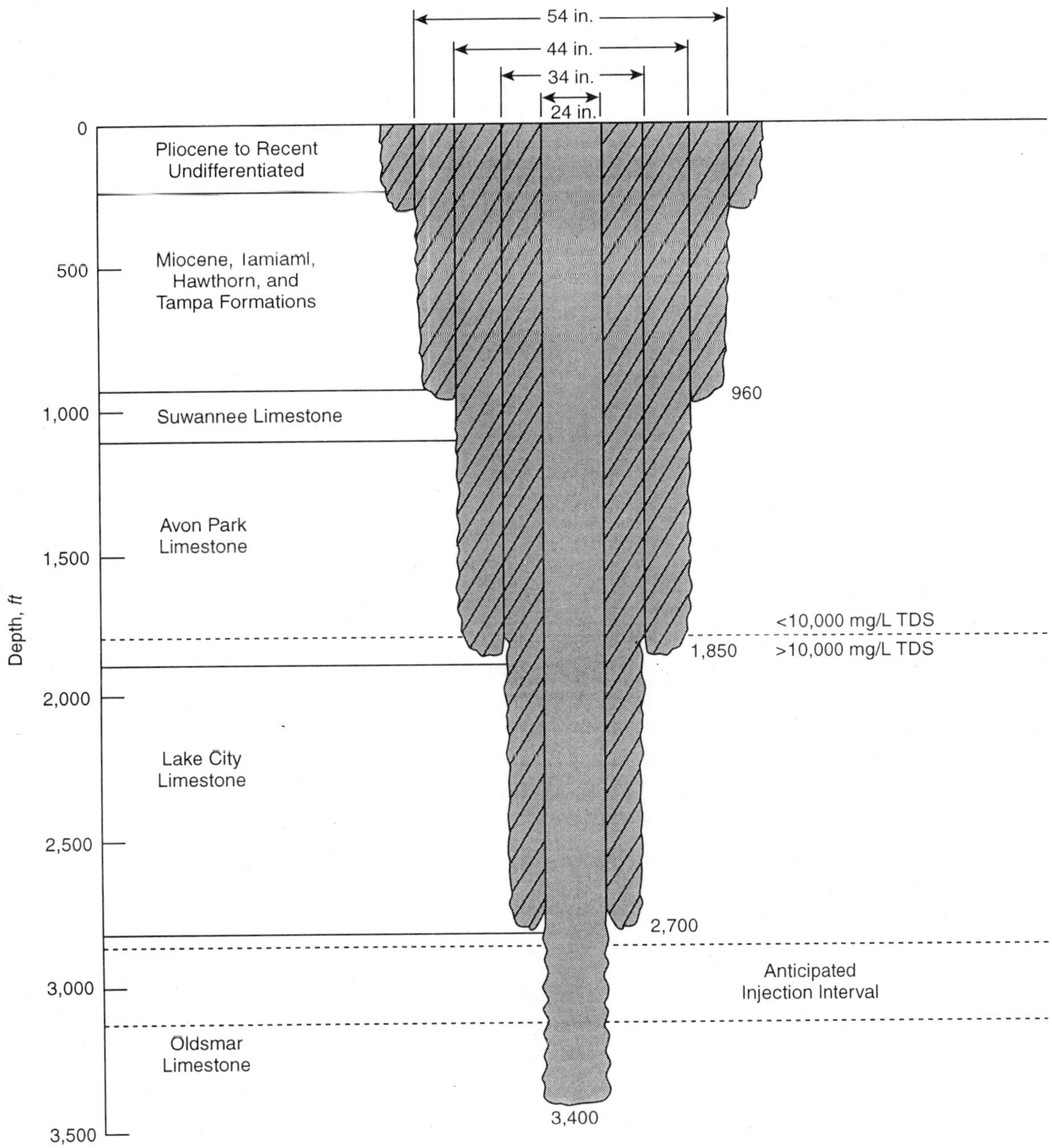

Figure 2-23 Deep injection well shaft

Figure 2-23 is a diagram of a deep injection well shaft, complete with casings and cement, for a typical south Florida injection well.

Casing. Deep injection wells are multicased, with the innermost casing set at the top of the injection zone. Three to four casings are generally used for injection wells in Florida. The depth of each casing depends on the geological environment surrounding the well. The main purpose of multiple, staged casings is to protect the upper freshwater zones from deeper, brackish zones and to reduce the possibility of fluid exchange between the different aquifers.

Cement grouting. The cement grouting surrounding each casing protects it from external corrosion, increases its strength, and prevents injected wastes from traveling to areas other than the designated injection zone. The type of cement and width of each cement layer surrounding a casing are regulated.

Injection zone. The location of a deep injection well is determined by the proximity of an acceptable injection zone. In order to avoid eventual plugging of the well, the water quality of the underground injection zone must be similar to the water quality of the membrane concentrate or the injection zone receiving water must be over 10,000 mg/L TDS (Conlon 1988). The injection zone should be characterized by high permeability and high transmissivity, which allow large volumes of concentrate to be injected without a large pressure buildup. The injection zone should also be located away from any abandoned wells, faults, or other hydrogeologic short circuits.

Florida has a distinctive underground environment that favors the use of deep injection wells. Five general injection zones exist within the state. The depths of these injection zones range from 700 to 8,000 ft below land surface, with the most widely used zone being a unique underground discharge zone located in southeastern Florida, referred to as the Boulder Zone. The water quality of this zone is similar to seawater. Thick, compacted layers of dolomite and limestone separate the saline water injection zone from overlying freshwater aquifers; this makes the Boulder Zone an ideal injection zone. Injection zone depths within the Boulder Zone range from 1,800 to 4,000 ft, whereas injection zone depths along Florida's west coast range from 700 to 1,600 ft. Florida's deepest injection wells are found in the Panhandle, with average depths ranging from 6,000 to 8,000 ft.

Pumping. The concentrate head is usually enough to move the concentrate to and down the injection well. If the concentrate head is insufficient, additional pumping will be required. The material of the injection well pump should also be compatible with the physical and chemical properties of the injected concentrate. Past experiences with injection systems have indicated that many difficulties are caused by improperly selected materials, resulting in corrosion of the injection pumps.

Storage. Temporary storage of the concentrate or an alternative method of disposal is needed to allow for maintenance and repairs of the well. Additionally, the well system may be shut down if monitoring systems and monitoring wells indicate leakage. The type of storage facility or stand-by disposal method is highly dependent on the location of the well and the conditions surrounding the well site. For example, if the injection well is located near the coast, a discharge canal or pipeline can be used to temporarily discharge the concentrate flow to a saline water body. A membrane facility located in Englewood, Fla., is two miles from the Gulf of Mexico and uses a previously existing concentrate disposal pipeline as a stand-by disposal system.

Regulatory monitoring. Following well system construction, detailed geophysical logs of the well site are required. Before well startup, television surveys of the well shaft and radioactive tracer surveys must be performed to validate the integrity of the well.

Accidental contamination of surface and subsurface fresh waters can be avoided by continuous monitoring of concentrate flow and wellhead pressure. Measurement readings of increasing pressure during steady operation can indicate possible clogging, whereas a sudden decrease in pressure is indicative of leaks within the casing, grout, or seal. Monthly testing of monitoring wells is also required by the Florida Department of Environmental Protection to ensure that the well system is not leaking into underground soils or water sources.

During general operations of the injection well, plugging, contamination, and wide variations in concentrate flow rates and pressures should all be avoided. Plugging can be caused by various occurrences, such as bacterial growth, suspended solid precipitation, or entrained air.

Irrigation

Using irrigation for concentrate removal can result in local water savings and help conserve natural resources by allowing aquifers to recharge.

History and regulations. Most literature concerning waste stream irrigation specifically addresses reclaimed municipal wastewater practices. Many of the concerns associated with wastewater effluent irrigation are the same as those associated with membrane concentrate irrigation.

Concentrate salinity and levels of other contaminants determine whether or not irrigation is a viable option. Many field studies and observations conducted in California were used to develop guidelines for evaluating irrigation water quality. These guidelines are based on concentrate salinity, toxicity, and soil permeability.

Salt is continually added to the soil with each irrigation water application, a practice that will eventually harm vegetation. The rate of saline accumulation depends on the quantity of salt applied and the rate at which it is removed from the soil by leaching. Adequate subsurface drainage is also necessary to avoid shallow water tables, which become an additional source of salts.

In addition to the effects of total salinity on vegetative growth and soil, individual ions can cause a reduction in vegetable growth as well. Toxicity caused by a specific ion occurs when that ion is taken up and accumulated by the vegetation. The ions of most concern in wastewater effluent irrigation are sodium, chloride, and boron.

The presence of salts in the soil reduces the rate at which water moves into the soil and also reduces aeration. If the soil permeability, or infiltration rate, is greatly reduced, then vegetation on the irrigation site cannot survive. Three factors affecting irrigation's long-term influence on soil permeability are the sodium content relative to calcium and magnesium, the carbonate and bicarbonate content, and the total salt concentration of the irrigation water.

If a membrane concentrate's salinity, toxicity, and soil permeability levels are determined suitable for irrigation, then the concentrate's ultimate effects on nearby surface water and groundwater must also be evaluated. Factors that affect an irrigation water's impact on groundwater include soil filtering, plant uptake, soil biochemical and physiochemical reactions, and soil particle adsorption. Nearby surface waters are affected mostly by the runoff generated from irrigation sites.

Other conditions must be met before concentrate irrigation can be considered as a practical disposal option. First, there must be a need for irrigation water within the vicinity of the membrane plant. A backup disposal or storage method must be available during periods of heavy rainfall. Monitoring wells need to be drilled before an operating permit can be obtained. At a membrane softening facility in Fort Myers, Fla., pilot-scale tests were conducted and conditions appeared favorable for the installation of a concentrate irrigation system. The resulting design plan specified a 150-acre golf course irrigation site with excess concentrate disposal to nearby stormwater systems.

Technical description. Irrigation with concentrate involves applying the waste stream to a vegetative surface and providing a runoff control system. The concentrate stream is applied to a vegetative area by a distribution system. There are three broad categories of distribution systems: (1) sprinkler or spray systems, (2) surface systems, and (3) drip systems. Drip systems have not been the conventional means of waste stream irrigation because the drip system emitters clog easily. However, these systems have been successfully used in California without plugging problems. Surface systems use narrow-graded (less than 15 ft) and wide-graded (100 ft or greater) borders or furrows for irrigation water distribution. Sprinkler systems use spray nozzles that move across the land. These spray systems can be used on variable soils, shallow soil profiles, rolling terrain, erodable soils, and areas where high water tables exist. Disadvantages of sprinkler systems include their higher initial capital cost, higher energy costs, mechanical failures, wind drift problems, and excessive evaporation losses.

The volume of runoff generated by an irrigation process often depends on the type of irrigation system used. Spray distribution systems do not generally cause a runoff of applied water, whereas surface systems produce some runoff. Ditches or drainage canals can be dug to retain runoff, or tailwater return systems can be used instead. A tailwater return system consists of a sump or reservoir, a pump or pumps, and a return pipeline. The pumps can be any convenient size, but a minimum capacity of 25 percent of the distribution system flow capacity is recommended.

Major cost parameters associated with a spray irrigation system include concentrate flow, distance to the irrigation site, transport pipeline costs, irrigation land purchase and preparation, distribution, pumping pressure, storage lagoons, subsurface drainage, and operation.

Land. Spray irrigation of concentrate is more land-intensive than other disposal methods. If an existing land area requiring irrigation, such as golf course, is not available, then land areas surrounding the plant must be purchased or leased for concentrate disposal. Land costs can fluctuate with the economy and vary depending on the location and characteristics of the site. Several options exist to purchase or control land used for a concentrate disposal system. Land may be purchased outright, leased on a long-term basis, or purchased and leased back to another party, for example, to a farmer. Purchasing land allows for complete control over the land and makes future expansion of the disposal site easier.

The soil must also be able to support a vegetative surface, which in most cases is a grassy surface. The need to prepare irrigation land by clearing or grubbing adds to overall disposal site costs and should be considered in selecting potential irrigation sites.

Spray irrigation systems also require land for service roads, buffer zones, storage lagoons, and equipment storage, in addition to the land area needed for the irrigation field. These additional land requirements should be taken into account in estimates of the total land requirements for the system.

Sprinkler system distribution. The predominant category of sprinkler systems, classified by their movement during application, is solid-set. Permanent solid-set sprinkler systems remain in one position during waste stream application. The major advantages of solid-set systems are low labor requirements and low maintenance costs. The main disadvantage is a high installation cost. A solid-set system consists of mainline and lateral pipes that cover the irrigation field with the sprinklers spaced along each lateral. Pumping to the distribution system is also needed to transport flow through the main pipeline to lateral pipelines and sprinkler heads.

Storage lagoons. Temporary storage facilities are necessary to retain concentrate during heavy rainfall periods or when an imbalance between concentrate supply and application occurs. The need for retention facilities is particularly important in Florida, where the average yearly rainfall approaches 52 in. in certain areas of the state.

Storage in a holding tank may be below ground in reinforced concrete structures or above ground in reinforced concrete or steel tanks. A storage facility can simply retain concentrate until disposal is again possible, or it can be designed to further facilitate concentrate removal. Storage in a reinforced concrete structure or steel tank merely retains concentrate, whereas storing concentrate in percolation ponds or storage lagoons further reduces concentrate volume via infiltration and percolation.

Subsurface drainage. Irrigation systems can be designed to include under-drainage or subsurface drainage. Subsurface drainage is necessary to provide a root zone area conducive to good vegetative growth. The presence of a high water table indicates poor subsurface drainage and demonstrates a need for installing subsurface drains. A high water table exists when the depth of the water table is less than 10 ft. The proximity of the irrigation site to canals, rivers, and other bodies of water should be considered when the irrigation site is chosen because seepage from other water bodies can contribute to subsurface drainage problems.

Subsurface drainage systems consist of a network of buried drainage pipes with open holes or perforations that recover effluent that has percolated through the soil. A collection basin is used to recover the water collected by the underdrains. This water can be reused by the irrigation system or possibly discharged into a receiving water body. USEPA cost-estimating curves have shown that including an underdrain system can increase the spray irrigation system capital costs by as much as 25 percent.

Operation. Operation and maintenance of a concentrate spray irrigation system is more labor-intensive than the other concentrate disposal methods. Labor requirements include sprinkler system repair and vegetative surface maintenance. Energy costs for pump operations also add to the system's total operational costs.

Rapid Infiltration Systems

Rapid infiltration basins (RIBs) have highly permeable soil bottoms that allow for rapid infiltration, percolation, and concentrate removal.

History and regulations. Treated wastewater effluent has successfully been disposed of by means of infiltration and percolation. Infiltration is defined as the entrance of water into the ground at the soil–water or soil–air interface. It is affected by the topography and surface conditions of the ground, the characteristics of the surface material, and the rate of precipitation. Percolation is defined as the flow of water within the soil in both the horizontal and vertical directions. It is controlled by gravity and by soil characteristics.

Rapid infiltration systems are capable of removing biodegradable organics, suspended solids, bacteria and viruses, phosphorus, nitrogen, and heavy metals. This particle removal occurs in the soil through filtration, adsorption, ion exchange, and biological action as the waste stream percolates through the soil.

Before a basin system can be constructed, extensive geological surveys are needed to determine soil permeability and background conditions. Rapid infiltration systems must also demonstrate that percolating concentrate meets water quality standards as well as primary and secondary drinking water standards. Monitoring wells are required to continually monitor groundwater quality surrounding the infiltration basin system.

Technical description. A rapid infiltration system consists of a series of basins commonly called *rapid infiltration basins* that are constructed with highly permeable soil bottoms. The concentrate stream exits the transport pipeline, enters the basin, quickly infiltrates through the porous surface soil, and then rapidly percolates into underlying soils. Examples of suitable soils include sands, sandy loams, and other coarse-textured soils.

Uniform application of the waste stream on disposal into the basins is necessary to avoid erosion. A simple splash block at the point of discharge may be used for small basins, whereas larger basins may require a more complex arrangement, such as pipes or troughs, to dampen the impact of the flow.

Hydraulic loading rate, which depends on infiltration, permeability, and hydraulic conductivity test results, is also a critical factor in RIB system design. The design loading rate must be based on the least permeable soil layer in the soil profile and on expected worst-case weather conditions. Concentrate discharge into the RIBs should be intermittent to maintain the design loading rate and soil capacity.

Major cost factors associated with rapid infiltration systems include the following: concentrate volume, distance from the membrane plant to the basin system, transport pipeline, land purchase and excavation, pumping pressure, and operational costs.

Land. Rapid infiltration systems, like spray irrigation systems, are land-intensive and often require a purchase of land. Preliminary rapid infiltration system site selection is based on a land area's soil and groundwater conditions, existing land use, flood potential, and proximity to the membrane plant. Site topography, soil type, and soil uniformity are also important because extensive cut-and-fill requirements can dramatically increase construction costs.

The total land area required for a rapid infiltration system is determined by the amount of land needed for transmission pipe easement, infiltration basins, access roads, pumping, buffer zones, a maintenance building, and future expansion.

Pumping. Pumping requirements for a rapid infiltration system are determined by the concentrate transport distance and concentrate head. The need for booster pumps along the distribution lines depends on the distance between basins.

Operation. Labor needs for system operation vary depending on whether the system is automated or manually operated. Infiltration basin maintenance involves raking the basin's surface to break up and loosen the soil surface. Basins may be raked following each drying period if labor and equipment are available.

Operation of a percolation system can be manually or automatically controlled. Automated operations involve time switches, float switches, tensiometers, and automatic valves. Regardless of whether a RIB system is manually or automatically controlled, operators should observe the rapid infiltration basins and make adjustments when necessary.

REFERENCES

Campbell, M.D., and J.H. Lehr. 1973. *Water Well Technology.* New York: McGraw-Hill Book Company.

Conlon, W.J. 1988. Disposal of Concentrate from Membrane Process Plants. In *Proc. AWWA Seminar on Membrane Processes.* Denver, Colo.: American Water Works Association.

Driscoll, F.D. 1986. *Groundwater and Wells.* St. Paul, Minn.: Johnson Division.

Konikow, L.F., and J.D. Bredehoeft. 1978. Computer Model of Two-Dimensional Solute Transport and Dispersion in Groundwater. Chapter C2 in *Techniques of Water Resources: Investigations of the US Geological Survey.* Reston, Va.: US Geological Survey.

Langelier, W.F. 1936. The Analytical Control of Anti-Corrosion Water Treatment. *Jour. AWWA*, 28(10):1500.

McDonald, M.G., and A.W. Harbaugh. 1988. *A Modular Three-Dimensional Finite-Difference Groundwater Flow Model.* US Geological Survey Techniques of Water Resources Investigations, Book 6. Reston, Va.: US Geological Survey.

Mickley, M., R. Hamilton, L. Gallegos, and J. Truesdall. 1993. *Membrane Concentrate Disposal.* Denver, Colo.: American Water Works Association.

Missimer, T.M. 1994. *Water Supply Development for Membrane Water Treatment Facilities.* Boca Raton, Fla.: Lewis Publishers.

Morin, O.J., and P.J. Malaxos. 1988. In *Proc. From Surface Water Discharge of Reverse Osmosis Concentrates.* National Water Supply Improvement Association. November 18–19.

National Water Well Association and the Plastics Pipe Institute. 1980. *Manual on the Selection and Installation of Thermoplastic Water Well Casing.* Worthington, Ohio: The National Water Well Association.

Reeves, M., D.S. Ward, N.O. Johns, and R.M. Cranwell. 1986. *Theory and Implementation for SWIFT II, the Sandia Waste-Isolation Flow and Transport Model for Fractured Media.* NUREG/CR-3328,

SAND 83-1159. Albuquerque, N.M.: Sandia National Laboratories.

Ridgway, H.F., C.A. Justice, C.W. Whittaker, D.G. Argo, and B.H. Olson. 1984. Biofilm Fouling of RO Membrane: Its Nature and Effect on Treatment of Water for Reuse. *Jour. AWWA*, 76(6):94–102.

Ridgway, H.F., M.G. Rigby, and D.G. Argo. 1985. Bacterial Adhesion and Fouling of Reverse Osmosis Membrane. *Jour. AWWA*, 77(7):97–106.

Sanford, W.E., and L.F. Konikow. 1985. *A Two-Constituent Solute Transport Model for Ground Water Having Variable Density.* US Geological Survey Water Resources Investigations Report 85-4279. Reston, Va.: US Geological Survey.

Taylor, J.S., S.J. Duranceau, W.M. Barrett, and J.F. Goigel. 1989. *Assessment of Potable Water Membrane Applications and Research Needs.* Denver, Colo.: American Water Works Association.

This page intentionally blank.

Chapter 3

Equipment and Installation

Reverse osmosis (RO) and nanofiltration (NF) systems require specialized equipment not found in conventional treatment facilities. This chapter presents a concise list of the components needed in each part of an RO or NF system, including source water delivery and pretreatment, membrane treatment, and posttreatment.

EQUIPMENT

Custom-engineered and standard prepackaged RO and NF treatment components are available from various membrane and equipment manufacturers (see appendix B for a list). Source water characteristics, hydraulic capacity, percent recovery, salt rejection, water flux decline, corrosion potential, and membrane type determine the necessary equipment specifications. Components include pretreatment equipment, the membrane process unit, permeate and concentrate posttreatment equipment, and auxiliary system equipment, such as equipment for membrane cleaning, instrumentation and control, and possibly drawback tanks for seawater RO systems. Figures 3-1, 3-2, and 3-3 show examples of different sizes of membrane units ready for installation.

Source Water

Source water comes from groundwater or surface water. The equipment listed in the following paragraphs may not apply to all systems, nor should the lists be considered exhaustive.

Surface water. Surface water delivery systems include the intake terminal, the intake pipe from the intake terminal to the intake chamber, trash racks, traveling screens, supply pumps, and the housing for the supply pumps, electrical equipment, and chemical feed equipment.

Pretreatment for surface waters may include conventional screening equipment, coagulation and flocculation components, sedimentation, and media filtration facilities, e.g., most seawater RO system intakes use in-line coagulation followed by

Figure 3-1 Photograph of a 5,500-gpd reverse osmosis system

media filtration. Economics and site conditions determine whether the media filters should be gravity-based, pressure-based, or a combination of the two types for two-step filtration. Disinfection, if warranted, may be performed. Chlorine residuals are usually removed by a dechlorination step before the membranes.

Groundwater. Groundwater source equipment includes pumping and electrical components. Pumps can be located in the well, at the top of the well, or on the ground near the well or wellfield. Groundwater usually does not require extensive pretreatment for the removal of suspended matter.

Further pretreatment for both groundwater and surface water before the membrane unit process can include

- acid pretreatment, which for sulfuric acid typically involves using positive-displacement metering pumps (or centrifugal pumps for large systems) drawing from a day tank
- scale inhibitor added by positive-displacement, chemical-metering pumps from a day tank
- cartridge filters containing multiple polypropylene-wound cartridges

Cartridge filters are 10 to 40 in. long and fit into stainless-steel or fiberglass-reinforced plastic (FRP) pressure vessels. The vessels may be oriented vertically or horizontally. Cartridges are rated for removal of particle sizes of 1, 2, 5, 10, and 25 microns, with the most frequently used sizes being 5 to 25 microns. The filter assemblies must include an inlet pressure gauge or a differential pressure gauge to

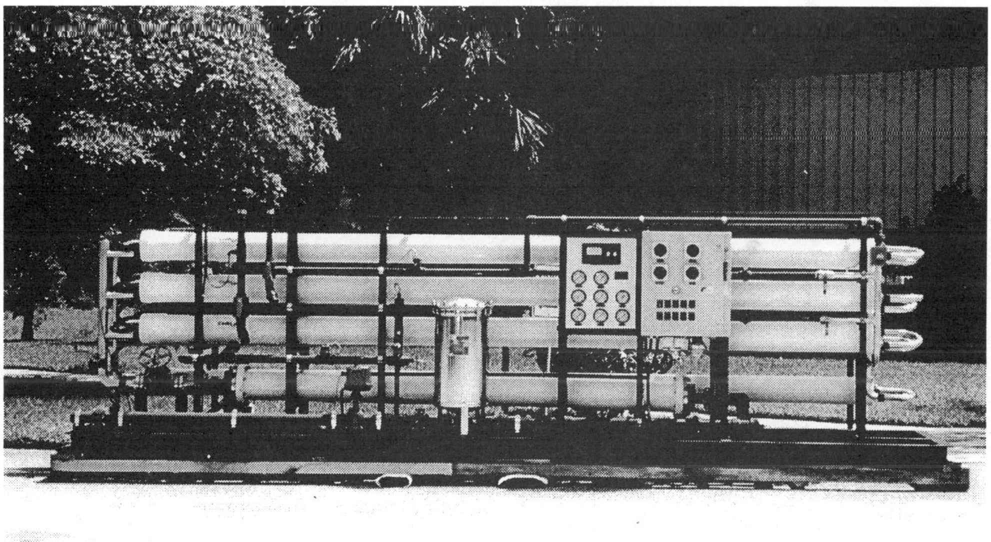

Figure 3-2 Photograph of a 115,000-gpd reverse osmosis system

Figure 3-3 Photograph of a 720,000-gpd reverse osmosis system

Figure 3-4 Cartridge filters and pumps at Acme Improvement District, Fla.

aid in determining when filter cartridges should be replaced. Figure 3-4 shows typical cartridge filter housings, high-pressure pumps, and valves.

Membrane Unit Process

The membrane unit is composed of RO or NF membrane elements, pressure vessels and support racks, source water or feedwater pumps, energy recovery devices (typically for seawater installations only), valves and piping, and instrumentation and controls (Figure 3-5). As discussed in chapter 2, system pressure often determines the types of components used in a membrane system.

Membrane elements and pressure vessels. Many types of RO and NF membranes are commercially available in spiral-wound and hollow fiber configurations. The membranes include those formulated with polyamide, polyethersulfone, polyvinyl alcohol derivatives, cellulose acetate blends, and other materials. Fiberglass-reinforced plastic and stainless steel are often used for pressure vessel modules. Membrane pressure vessels and modules are usually installed on welded steel or fiberglass-reinforced plastic support racks.

High-pressure pumps and energy recovery equipment. Horizontal centrifugal and vertical turbine can-type pumps are typically used for RO and NF systems. Smaller-capacity seawater RO systems often use positive-displacement pumps with suction and discharge accumulators or pulsation dampeners to minimize pressure fluctuations. Variable-speed pumps are commonly used in small systems that do not employ energy recovery systems, although seawater systems may employ both variable-speed pumps and energy recovery. Larger systems typically use centrifugal pumps.

Seawater RO and some brackish RO systems employ energy recovery equipment. Energy is recovered by using the residual high pressures present in the

Figure 3-5 Reverse osmosis membranes (three trains of 4 mgd each) at City of Plantation, Fla.

concentrate water stream. Common energy recovery devices include integrated reverse-running centrifugal pumps and impulse turbines coupled to the feedwater pump motors and hydraulic turbochargers to boost pressure to the second stage. The construction material for wetted parts of feedwater pumps and energy recovery equipment is normally type 316 stainless steel.

Valves. A variety of valves are commonly used in a membrane unit process. The type of material used for valves depends on the required pressure ratings and the corrosivity of the contacting liquid. Typically, polyvinyl chloride (PVC) valves are used for low-pressure systems and stainless-steel valves for high-pressure applications.

Valves designed for a high-pressure drop are used for concentrate throttling to minimize cavitation and operational noise. Butterfly and ball valves are used to isolate individual manifolds and membrane trains. Check valves are commonly used on pump discharge lines and on concentrate lines from each membrane train. Pressure relief valves are typically included on the permeate stream between the membranes and any isolation valves, depending on the system design and stack array configuration. In some instances the wellfield may be controlled via diaphragm-actuated hydraulic control valves. Most systems employ multiple sample valves located in the membrane array for water quality monitoring purposes.

Piping. High-salinity brackish water systems typically use type L stainless steel for piping. Nanofiltration or low-salinity RO systems use type 304 stainless steel and, in some instances, PVC or FRP piping.

Instrumentation and control. Instrumentation and control systems can be as basic as manual control with automatic shutdown features for pump and membrane protection, or as complicated as a supervisory control and data acquisition (SCADA) system. SCADA systems are often based on programmable logic controllers

Figure 3-6 Person-machine interface for a supervisory control and data acquisition system

and remote telemetry units (RTUs) that are supervised by a host computer located in a control room near the membrane skids and high-pressure pumps. The use of personal computers or industrial grade, person–machine interfaces allows a compact control room installation, as shown in Figure 3-6.

Systems designed for automatic control can monitor chemical feed systems, and they have alarm and report generation capabilities. Various interlocks can be provided to monitor pressure, flow, transfer pumps, scrubbers, and ancillary equipment. Other interlocks also protect the system from operating outside acceptable limits. An in-line field turbidimeter can be integrated into the system to monitor pretreated membrane feedwater. The main switchboard and motor control centers provide power to major equipment, power panels, and membrane feed pumps. Modern electronic sensors are often used to avoid the limitations of analog mechanical recording equipment.

The main switchboard and motor control centers provide power to major equipment, power panels, and membrane feed pumps. Distribution panels contain circuit breakers for low-electrical-consumption devices, such as room lights, instrumentation equipment, small motors, and other equipment. Certain equipment can also be controlled from the membrane process area via discrete local control panels, as shown in Figure 3-7.

Posttreatment

As discussed in chapter 2, posttreatment of both permeate and concentrate varies depending on specific applications and regulatory requirements. Permeate and, in some cases, concentrate posttreatment processes, generally include a PVC or FRP degasifier with a scrubber for removing carbon dioxide and hydrogen sulfide, as well

Figure 3-7 Local control panel at a membrane process area

as chemical addition for stabilization. Chlorination and dechlorination of concentrate may also be required in addition to the use of diffusers to mix concentrate with the receiving water body. If deep well injection is used for concentrate disposal, then appropriate piping, valves, and pumps are required.

Membrane-Cleaning System

Typically, membrane-cleaning systems contain a nonmetallic storage-feed tank and a low-pressure stainless-steel or plastic pump with valves and piping. Cleaning systems can be modified to provide heated cleaning solutions, separate solution and feed tanks, a cartridge filter, a mixer, and instrumentation and control. Cleaning solutions are commonly transported to and from membrane modules through PVC piping and hoses. Spent cleaning solution discharges to an FRP tank, where the pH and water quality can be adjusted before the solution is released to a sewer, deep well, or other disposal system.

Ancillary Equipment

Ancillary equipment may include emergency generators, elevated and ground storage tanks, disinfection equipment, laboratory equipment, and high-service pumps. Seawater RO systems usually have a nonmetallic drawback tank that quickly provides permeate water to the membrane modules during system depressurization on shutdown.

COSTING

Costs for both construction and operation of membrane processes are dependent on site-specific factors. Plants of the same capacity and process type can therefore have very different construction costs. Site-specific factors affecting construction costs include the following:

- plant size or capacity
- blending of source water with permeate
- plant components included in the cost
- concentration of total dissolved solids (TDS) and other constituents of the source water requiring removal
- concentrate disposal
- intake type for source water
- pre- and posttreatment requirements
- indirect costs
- other costs

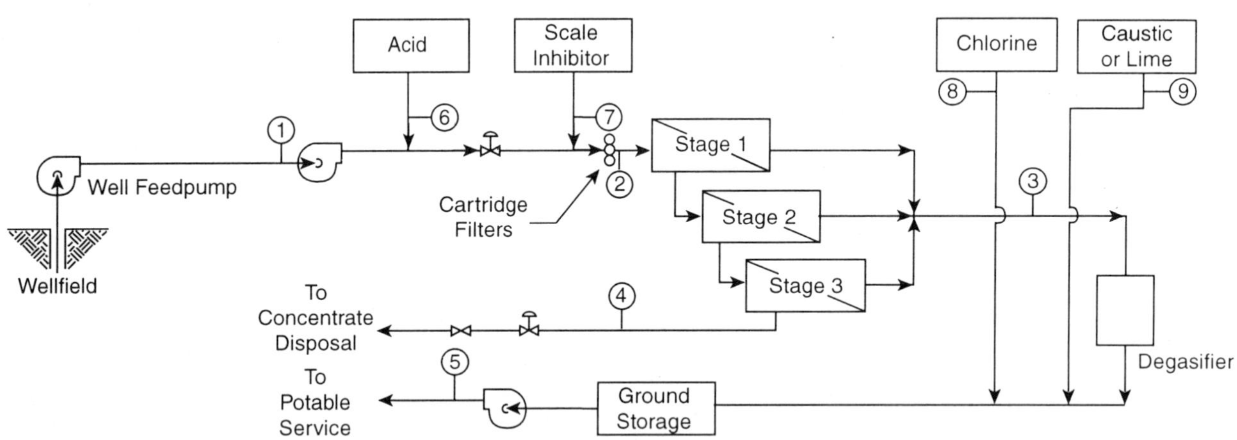

Process Stream	①	②	③	④	⑤	⑥	⑦	⑧	⑨
Process Conditions	Feedwater	Feedwater to Membranes	Product Water	Concentrate	Finished Water	Acid Treatment, *mg/L*	Scale Inhibitor Treatment, *mg/L*	Chlorine Treatment, *mg/L*	Caustic Lime Treatment, *mg/L*
Flow, *gpm*	868	868	694	174	674	100	4.0	2.0	37.5
Pressure, *psig*	50	285	10	246	100	--	--	--	--
TDS, *mg/L*	2,000	2,000	110	9,641	194	--	--	--	--

Figure 3-8 Process flow schematic for the reference design

Table 3-1 Reference design parameters

Parameter	Value or Type
Water supply source	Groundwater
Source water TDS, *mg/L*	2,000
Permeate TDS, *mg/L*	110
Permeate quantity, *mgd*	1.0
Finished water quantity, *mgd*	1.0
Finished water TDS, *mg/L*	194
Number of trains or arrays	1
Number of stages	3
Number of membranes	168
Number of source water pumps	1
Electrical requirement, *kW*	136
Process recovery, *%*	80
Product water flux, gpd/ft^2	18
Concentrate disposal	To surface water body

Reference Design

To determine how costs are affected by these site-specific factors, a reference design for a typical brackish water, low-pressure RO membrane system treating groundwater is included in this section. Cost changes can be determined based on the reference design. Figure 3-8 provides a process flow sketch for this reference design. Table 3-1 lists design parameters, and Table 3-2 gives the capital or construction cost information. In this chapter's discussion, all costs presented are in January 1994 dollars represented by an *Engineering News Record* construction cost index value of 5,336.

A review of this information reveals that the following site-specific factors have been included in the reference design:

- Plant size is 1.0 mgd.
- No blending of source water occurs.
- Plant components include
  - wellfield
  - pretreatment and posttreatment
  - RO skids with cartridge filters and pumps
  - product storage and high-service pumping
- TDS of the feedwater is 2,000 mg/L.
- Concentrate disposal is to a surface body of water.

The following paragraphs address differences in capital costs for changes in these site-specific factors.

Table 3-2 Construction cost for the reference design

Cost Item		Value, *$1,000*
Direct construction costs		
Site development		25
Source water supply		191
Pretreatment and cleaning		74
Process equipment cost		742
Reject disposal		35
Building and auxiliaries		170
Posttreatment, storage, and pumping		472
	Subtotal direct cost	1,709
Indirect construction costs		
Freight, insurance, taxes		68
Construction overhead		308
Owners' costs		137
Contingency		171
	Subtotal indirect cost	684
	Total Construction Cost	2,393
Other costs		
Working capital		490
Interest during construction		256
	Subtotal other cost	746
	Grand Total	3,139
	Unit construction cost	\$3.14/gpd

Plant Capacity

The required plant capacity, in conjunction with the recovery factor, determines the required feedwater capacity and establishes the cost of the membrane process, source water supply, and pretreatment and posttreatment systems. The capacity determines the number of membranes required, and the recovery factor defines the size of the auxiliary equipment. Once this cost has been established for a particular recovery rate, the cost for a plant of different capacity may be estimated from the following expression:

$$C_B = C_A\left(\frac{S_B}{S_A}\right)^n \tag{3-1}$$

Where:

C_B = cost of the new size plant, in US dollars
C_A = cost of known plant, in US dollars
S_B = capacity of new plant, in gpd
S_A = capacity of known plant, in gpd
n = scaling factor

For a brackish water process, a scaling factor of 0.80 may be used to determine costs of different-sized plants. As an example, assume a 1.0-mgd plant capacity has a total construction cost of $2,400,000. Based on Eq 3-1, the construction cost of a 5.0-mgd plant would be estimated at $8,697,000. Note that the unit cost has been reduced from $2.40/gpd for the 1.0-mgd plant to $1.74/gpd for the 5.0-mgd plant. This decrease in unit costs with increasing plant capacity is referred to as *economy of scale*.

Economy of scale can be seen in Figure 3-9, where the unit costs for process equipment alone are plotted for plant capacities between 1 and 20 mgd. This information indicates that the unit cost for a 20-mgd plant size is roughly two-thirds the unit cost of a 1.0-mgd plant size.

Recovery

The plant recovery factor affects the cost of auxiliary equipment, such as wellfields, pumping, and piping and valves; it also affects the cost of concentrate disposal. Membrane plants are always designed for the highest recovery possible because this will keep the auxiliary equipment cost at the lowest level. The maximum recovery for an RO system is limited to about 90 percent.

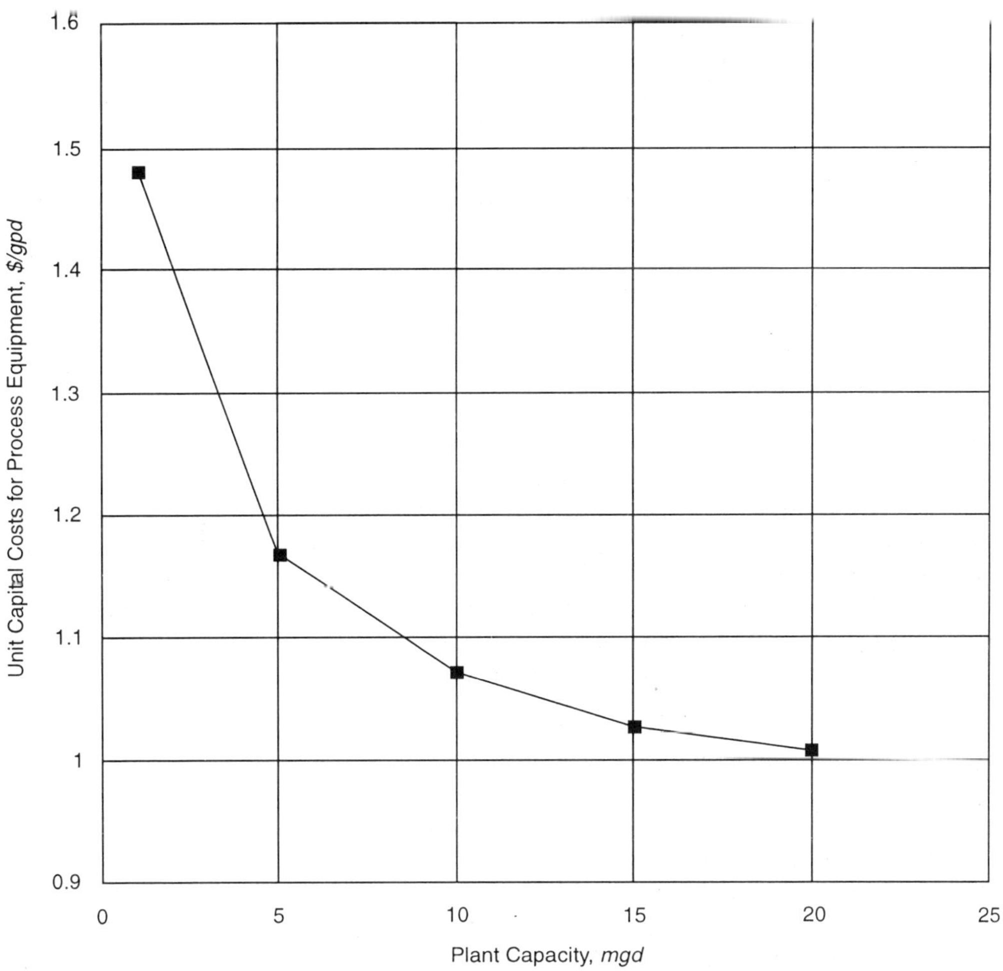

Figure 3-9 Economy of scale for reverse osmosis plant process equipment

The capital cost given for the reference design was determined at a recovery of 80 percent. For this recovery, the cost for source water supply and pretreatment equipment is about 22 percent of the total. For a design using a 70 percent recovery and the same number of membranes, a new cost for the feedwater system may be obtained from Eq 3-1 with a scaling factor of 0.65 rather than 0.80. The difference in scaling factor is due to the type of equipment under consideration, mainly pumps and piping for the feedwater system. The feedwater flow rate for the reference design was 868 gpm, and the flow rate for the case of 70 percent recovery would be 992 gpm. Using these new figures results in an estimated plant cost of $2,435,000, which is a cost increase of 2 percent relative to the higher recovery reference design.

Blending

It may be possible for membrane plants to blend some of the source water with permeate from the plant. For many brackish water applications, permeate from the membrane process is relatively low in TDS compared to potable water requirements. The advantage gained by blending is a reduction in required plant size, which results in reduced capital and operating costs.

For example, the reference design RO plant treating a water supply of 2,000 mg/L TDS may have a permeate quality of 110 mg/L TDS. Because this figure is much lower than the Safe Drinking Water Act (SDWA) guideline of 500 mg/L TDS, source water can be mixed or blended with the permeate to result in a finished water quality of 500 mg/L TDS. The mass balance equation used to determine the blending process is:

$$Q_F\, C_F = Q_P\, C_P + Q_b C_b \tag{3-2}$$

Where:

Q_F = finished water flow rate
Q_P = permeate flow rate
Q_b = bypass flow rate
C_F = concentration of finished water (500 mg/L for present example)
C_P = concentration of permeate (110 mg/L for present example)
C_b = concentration of bypass water (2,000 mg/L for present example)

It turns out that, for the assumed 1.0-mgd case, approximately 20 percent of the source water can serve as bypass flow to be mixed with the permeate. Consequently, the capacity of the RO plant can be reduced by 20 percent. Equation 3-1 shows that the capital cost savings is about $392,000, about a 16 percent reduction in capital cost.

The blending of source water does, however, have specific limits. Constituents that could result in limitations on the amount of blending might include

- trihalomethane concentration in finished water
- hardness of finished water
- concentrations of individual constituents in the source water that, when blended with permeate, could still exceed the SDWA regulated value

Components Included in the Cost

Cost estimates given for constructing membrane plants do not always include all the equipment necessary for a complete installation. For example, costs given for

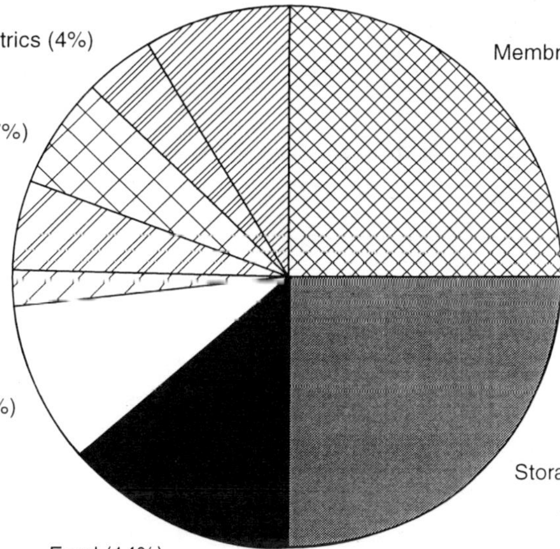

Figure 3-10 Construction cost breakdown for a brackish reverse osmosis or nanofiltration process

membrane systems frequently do not include product storage and high-service pumping, or they may exclude the water source, such as wells, or the intake structure.

Figure 3-10 gives the breakdown of the components required for a brackish water RO or NF membrane system installation. Percentages given are for the reference design. Review of this figure shows that if, for example, the source water supply or storage and pumping were omitted from the cost estimate, approximately 14 or 25 percent of the cost would be overlooked, respectively. The membrane system equipment itself represents only a portion of the total cost, because it excludes pretreatment and posttreatment, buildings, feedwater supply system, product storage, and high-service pumping.

Source Water Quality

The term *source water quality* refers to the amount of inorganic and organic material in the source water to be treated in the membrane plant. These factors will determine

- recovery factor
- blending quality
- electrical consumption
- pretreatment and posttreatment requirements

Recovery factor and blending were discussed previously; the following paragraphs are devoted to the last two items.

The amount of electricity consumed is in direct proportion to the concentration of inorganics in the supply. For membrane systems, the greater the concentration of inorganics, the greater the osmotic pressure to overcome. Greater pressure requirements result in a need for increased pump horsepower. A comparison of specific energy use measured in kilowatt-hours per 1,000 gal (kWh • kgal) of water production is given in Figure 3-11. This figure plots the reference design process energy requirements for

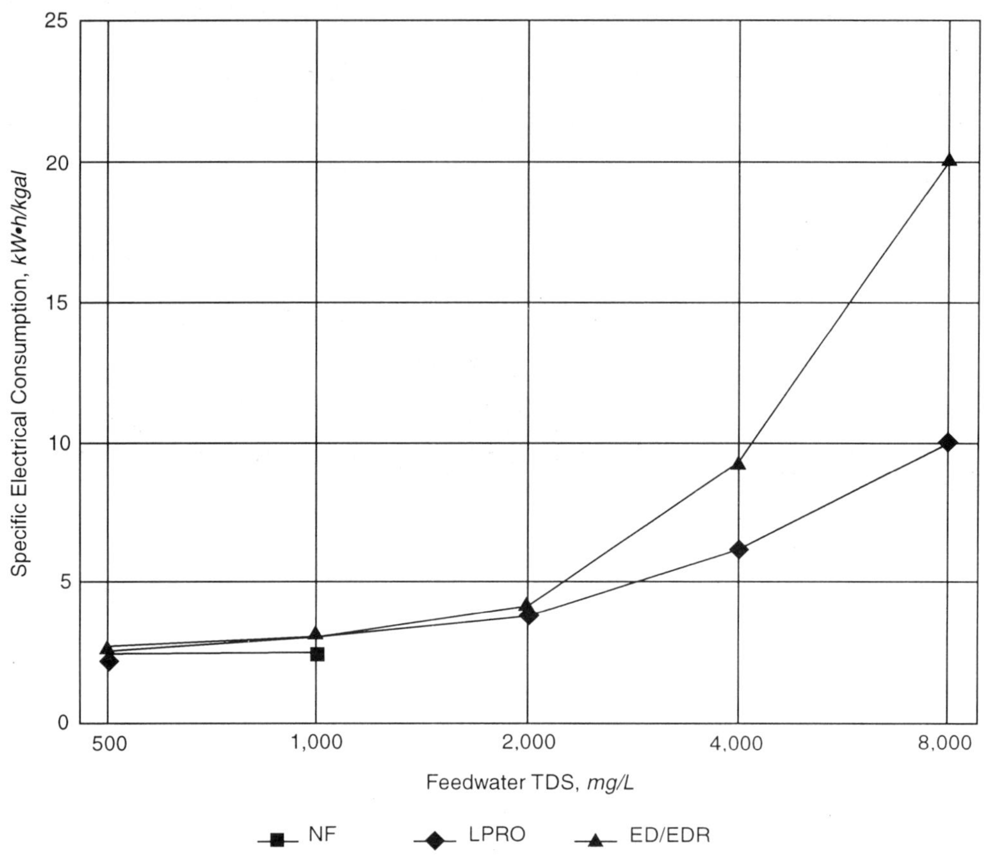

Figure 3-11 Approximate specific power consumption for nanofiltration, reverse osmosis, and electrodialysis/electrodialysis reversal processes

an NF membrane, a low-pressure RO (LPRO) membrane, and an electrodialysis/electrodialysis reversal process. Note that an LPRO membrane represents newer generations of membranes capable of achieving significant chloride reduction at relatively low pressures (around 200 psi). Figure 3-11 indicates that as TDS levels increase, energy requirements increase for all processes. Also, for the lowest TDS concentrations, all membrane processes have similar energy requirements.

The amounts or concentrations of organics in the supply have a distinct effect on the RO and NF processes. For membrane processes, organics cause an accelerated fouling of the membranes. This leads to frequent cleaning, which in turn leads to quicker deterioration of the membranes. To counter the fouling problem in membrane systems, the design flux is reduced. Typically, surface water supplies or shallow groundwater supplies have high organics concentrations, whereas deep groundwater supplies have low organics concentrations.

In the reference design, the flux is 18 gpd/ft^2, a typical rate for treating a groundwater supply. This flux requires using 168 membranes of 8-in. diameter. Each membrane has a surface area of about 330 ft^2. Had this supply been assumed to be high in organics, the design flux would have been lowered to the 10–12 gpd/ft^2 range.

Assuming a flux of 10 gpd/ft^2, the number of membranes would have increased to 302, resulting in an additional total capital cost of approximately $550,000.

Concentrate Disposal

As discussed in chapter 2, concentrate can be disposed of in one of the following ways:

- saline surface waters, including brackish waters and seawaters
- solar ponds
- sewer systems
- deep well injection
- further concentration to dryness, as in zero discharge

To compare disposal costs and evaluate the effect recovery can have on these costs, estimates were prepared for the reference design for recoveries of 80 and 90 percent. These results are given in Table 3-3, which shows the wide variation in capital costs for each type of disposal system. The least expensive option, and the one primarily used for membrane treatment plants, is disposal to a brackish water source. A large investment in capital is required for building solar evaporation ponds, the most costly option. Other systems fall in between these two extremes, and costs for systems depend heavily on such other site-specific cost items as land and sewer use. Costs for disposal systems at an assumed recovery of 90 percent indicate the significant construction cost savings that can be realized at higher recoveries.

Not shown in Table 3-3 are the costs to operate the disposal systems. All systems, with the exception of thermal evaporation, do not require significant amounts of electricity or chemicals for normal operation. Thermal systems,

Table 3-3 Costs for different concentrate disposal methods (based on reference design)*

	Cost, *dollars*		
Disposal Method	80 Percent Recovery	90 Percent Recovery	Reduction, *percent*
Surface water	34,000	–	–
Solar ponds	4,800,000	2,100,000	56
Sewer system	625,000	277,500	56
Deep well injection	2,119,500	2,119,500	0
Thermal evaporation (zero discharge)	3,599,900	1,413,000	61

*Assumptions are as follows:

Brackish water body	Location (distance from membrane plant), 2,000 ft
Solar ponds	Evaporation rate, 60 in. per year Land cost, $100,000/acre Design life, 30 years
Sewer system	Capital cost charge, $2.50/gpd Operations and maintenance (use) charge, $400/mgd
Deep well injection	Well diameter, 12 in. Well depth, 3,000 ft Costs include monitoring well and owners' costs.
Zero discharge	Vapor compression unit capital cost, $9.50/gpd Crystallizer unit capital cost, $25/gpd

however, require both electricity and chemical treatment and may require steam as well. Operating costs play an important role in evaluating and selecting the optimal disposal method.

Source Water

The reference design assumes the use of wells for groundwater supply. Usually, the construction cost of a surface water supply will be greater than that for a groundwater supply. A surface water supply requires a larger number of components for intake construction, as previously discussed in this chapter. Furnishing the reference design with a surface water intake system is estimated to cost $690,000. This is considerably more than the cost for the groundwater supply of the reference design of $191,000 (see Table 3-2).

Pretreatment and Posttreatment

The reference design includes minimal pretreatment consisting of

- acid feed
- sodium hexametaphosphate
- cartridge filtration

Only minimal pretreatment has been included because, for the well supply, the ground strata perform the filtration step. Thus, the concentration of colloids is low, and silt density readings of 5 or lower are typical.

For RO systems treating surface waters, some form of filtration is required to reduce colloidal content. Additional chemical treatment usually accompanies this filtration. At a minimum, dual-media or mixed-media filtration is required. For some installations, however, where colloidal content is excessive, coarse filtration will be required ahead of the other filtration step. Filtering a surface water supply can be capital intensive (Table 3-4).

The capital cost for posttreatment depends on permeate quality and whether or not source water is blended with permeate. Typical finished water quality goals are shown in Table 3-5.

Table 3-6 shows the permeate quality that can be expected from the reference design after the degasifier. This water is extremely corrosive and must be treated before being pumped to the distribution system. Chemical treatment, blending, or a combination of these two methods can be used.

Chemical treatment involves adding lime (usually quicklime), caustic, or a combination of the two. For this particular example, a combination of lime and caustic was used to achieve the final finished water quality goals. Feed rates required for the 1.0-mgd reference design included lime at 37.5 mg/L, or 312.0 lb/d, and caustic at 15 mg/L, or 125.2 lb/d.

The cost for the chemical treatment system, pumps, piping, and valves (excluding bulk storage) is $277,100. In comparison, blending feedwater with a membrane-treated water in a 1:4 ratio meets the finished water quality goals (Table 3-7) without the need for chemical treatment.

Indirect Costs

Once the installed construction costs have been determined, the indirect costs should be estimated, including

- freight, insurance, taxes, and duties (normally 4 to 6 percent of direct costs)
- construction overhead (contractor engineering, procurement, field supervision, and general overhead costs)

Table 3-4 Cost estimate for filtration pretreatment for reference design using surface water source

Component	Direct Construction Cost, *dollars*
Filter structure	328,500
Filter media cost	13,300
Air backwash system	132,600
Total	474,400

Table 3-5 Typical finished water quality goals

Constituent	Goal
Alkalinity (as $CaCO_3$)	40 mg/L or greater
Hardness (as $CaCO_3$)	40 mg/L or greater
Langelier Saturation index	Near zero or positive
pH	7.0 to 8.5, or as required to satisfy LSI goals
TDS	Less than 500 mg/L

Table 3-6 Typical permeate quality for the reference design (after the degasifier)

Constituent	Value
Alkalinity (as $CaCO_3$), *mg/L*	15.50
Hardness (as $CaCO_3$), *mg/L*	23.70
Langelier Saturation index	−2.16
pH	7.26
TDS, *mg/L*	110.00

Table 3-7 Typical posttreatment quality for the reference design

		Finished Water		
Constituents	Permeate	Chemical Treatment	Blending	Water Quality Goal
Alkalinity (as $CaCO_3$), *mg/L*	15.5	65.0	48.2	40+
Hardness (as $CaCO_3$), *mg/L*	23.7	86.0	119.4	40+
Langelier Saturation index	−2.16	0.12	−0.47	0
pH	7.26	7.88	7.75	7.0–8.5
TDS, *mg/L*	110	194.3	480.6	<500

- contingency
- owner costs (engineering design, bond insurance costs, working capital, interest during construction)

Engineering fees for plant design vary considerably depending on the magnitude of the project and the scope of services. In terms of a percentage of direct costs, the engineering fee for detailed design of the reference plant would exceed 15 percent.

Raising capital for construction is commonly required, but not always. Thus, capital-raising costs generally do not become part of the project estimate. Typical bond issuance costs, if required, include the following:

- bond finance underwriter (10 percent of direct capital costs)
- bond legal and issue printing costs (5 percent of direct capital costs)
- bond insurance (1 percent of the total bond debt)

Working capital includes funds needed during construction and initial operation. This item is usually taken as 50 to 60 percent of the first year's operations and maintenance cost. For the reference design, this amounts to $490,000. Interest during construction is due during the construction period before plant operation. If it is assumed that it will take 14 months to construct the reference design, the interest will be $256,000.

Summary

Table 3-8 summarizes the cost impacts discussed previously. This summary shows the dramatic cost impacts of the site-specific factors studied. Construction costs could be as low as about $2 million or as high as more than $7 million for the same plant. The cases shown assumed site-specific impacts on the basis of single occurrences; however, these impacts can be combined. For example, for a surface water supply, a low-flux design, extensive pretreatment, and a surface water intake structure would probably be required. These would combine to add $4,254,500 to the reference design capital cost, for a total cost of $6,650,500.

Table 3-8 Summary of cost impacts relative to the reference design

Item	Cost Impact, *dollars*	New Plant Cost, *dollars*	Percent Change Relative to Reference Design Cost
Reference design	–	2,396,000	–
Recovery at 70 percent	+39,000	2,435,000	+2
Blending at 20 percent	–392,000	2,004,000	–16
No well supply	–267,300	2,128,700	–11
Low flux (10 gpd/ft^2)	+550,000	2,946,000	+23
Concentrate disposal			
Solar ponds	+4,800,000	7,196,000	+300
Sewer system	+625,000	3,021,000	+26
Deep well	+2,119,500	4,515,500	+89
Zero discharge	+3,599,900	5,995,900	+250
Surface water intake	+659,400	3,055,400	+27
Extensive pretreatment	+663,900	3,059,900	+28
Chemical posttreatment	+277,100	2,673,100	+12
Bond cost requirements	+423,300	2,819,300	+18

Figure 3-12 Photograph of a 330,000-gpd, skid-mounted reverse osmosis system

INSTALLATION

Depending on the size of the unit, RO or NF equipment may be furnished in a skid-mounted configuration (Figure 3-12) or prefabricated for assembly on-site. Many system manufacturers offer packaged skid-mounted units of varying capacities. A typical packaged unit consists of a cartridge filter, RO or NF feed pump, membrane modules, instruments and controls, instrument panel, motor control center, and all interconnecting piping and valves on a skid. Acid and antiscalant feed systems for pretreatment are offered on separate skids. However, for plant capacities less than 100,000 gpd, chemical feed systems may be installed on the membrane module skid.

Site conditions and aesthetic reasons determine whether the plant should be installed indoors or outdoors. Extreme weather conditions or a corrosive environment may require the plant to be installed indoors, and protection against freezing by insulation and heat tracing may be necessary. Where outdoor installation is preferred, the plant should be installed on a concrete slab and the module bank protected from direct sunlight. A small outdoor membrane installation is shown in Figure 3-13.

Provision should be made for floor drains near the module banks. Product tube probing, membrane element replacement, and cleaning procedures all involve spilling large volumes of water, requiring adequate drainage facilities. In large installations, pipe trenches connected to the plant drain system serve as drains for spillage from the membrane modules.

Reinstallation

Adequate clearance at both ends of the membrane module must be provided for loading and unloading membranes. Clearance should be at least equal to the length of the

Figure 3-13 Outdoor installation of a membrane system

element plus a working space of 12 to 18 in. For example, a typical 40-in.-long, spiral-wound membrane element would require at least a 52-in. clearance. Feedwater, concentrate, and permeate headers and their connections to the pressure vessels should be designed to facilitate easy access for pressure vessel probing. The feedwater piping should be designed to allow debris to be cleaned up before system startup. Blowoff circuits, such as tee fittings with valves and piping to the ground surface, should be installed at certain strategic locations to allow the operator to run wells to waste.

The chemical feed system should be designed and installed to permit drawdown to as low as 1 in. above the suction pipe connection on the day tank. The installation should allow easy access for changing hydraulic oil where required without the need to disconnect pumps or piping. Also, drain connections should be suitably sized and piped for safe draining of the tanks. Where possible, facilities to safely flush chemical pumps (excluding concentrated acid feed pumps) with clean water should be provided. The piping system should be free from dead-end sections, which may harbor microorganism growth.

Generally speaking, no restrictions should be placed in the permeate line from the module bank to the storage tank or degasifier. If isolation valves are necessary, protection should be installed so that the module cannot be brought on-line without these valves first being opened. Most membrane manufacturers recommend that permeate pressure never exceed the feedwater or concentrate pressure. Pressure from the permeate side can dislodge the membrane from the backing material or weaken glue lines in the case of spiral-wound membrane elements. If permeate lines contain isolation valves that can completely prevent permeate flow (deadheaded), pressure relief protection must be installed.

Membranes and pressure vessel housings should be protected from extreme conditions, such as direct sunlight, steam lines, freezing, and heater exhaust. Trenches inside and outside the building must be covered in traffic areas. In other areas where trenches need to be accessed, handrails should be constructed for worker safety.

The erection and assembly instructions provided here are to assist the reader in understanding general guidelines and are not intended to be a substitute for the experience of the persons assigned to assembling or installing equipment. To protect the equipment and keep the warranty valid, all equipment must be stored, handled, adjusted, and maintained according to instructions provided by the manufacturer.

The following conditions are assumed for this discussion of installation:

- all civil engineering work has been completed
- the electric utility has provided power
- the wellfield and well pumps are active and available for use
- arrangements have been made for the disposal of flush water, chlorinated water, and wastewater
- means for the disposal of concentrate have been provided

Receiving Materials

Because equipment arrives before installation begins, a qualified person should be responsible for receiving the equipment and storing it according to manufacturers' recommendations. On arrival, equipment should be inspected for damage and checked against the material lists supplied by the equipment manufacturer or supplier. Instruments, electronic devices, electric motors, motor control centers, and instrument panels may have to be stored in a relatively cool, dry, and secure environment. All rotating equipment should be kept accessible to allow the shafts to be moved manually at least once a week. Membrane elements should be stored in their original bags and boxes according to the manufacturer's recommendations. In extreme climates, temperature-controlled storage may be required. Preassembled piping, support frames, and other similar equipment should be properly supported to maintain their shape. This is also important for the plastic piping supplied in straight lengths. All open ends should be closed to keep dirt and debris out of the piping.

The membrane system furnished by the manufacturer may have several components shipped separately for assembly and installation on-site. Typically, each carton will include the supplier's itemized packing list of parts. Items are tagged with corresponding material list numbers. This list should be checked against the items received and then retained for future reference. Any missing or damaged items should immediately be reported to the supplier for appropriate action. If possible, the problem should be noted on the delivery freight bill or other delivery data.

Skid-mounted equipment should be moved by the lifting lugs. If there are no lifting lugs, a forklift may be used to pick up the equipment from under the skid.

Installation Guidelines

Large and heavy equipment is set up first. Pumps, motors, control panels, transformers, storage and cleaning tanks, and any equipment that requires the use of a crane should be placed first to free up the crane as soon as possible. If the membrane unit does not arrive mounted on a skid, the next step is the erection of the

RO or NF frames. After the frames have been secured, the pressure vessels should be mounted on the frames and strapped down. Feedwater, concentrate, and permeate piping should be installed and supported as soon as possible.

Once piping and support installation is completed in any area, electricians can install conduits, cable trays, local control panels, and instrument panels. Instrument tubing and sample lines should be installed after the majority of the equipment is in place.

Manufacturers' recommendations should be followed to install RO or NF feed pumps, chemical-metering pumps, and instruments. Temporary pipe supports provided during the assembly of interconnecting piping should be replaced with permanent supports as soon as possible.

Centrifugal pump installation. The Hydraulic Institute provides comprehensive instructions for installing almost any type of pump. An excerpt of guidelines provided by the Hydraulic Institute is reproduced below:

1. The pump should be located so that a short, direct discharge pipe with the least number of elbows and fittings may be used to minimize head loss from friction. If practical, place the pump so that it will be accessible for inspection during operation.

2. The foundation should be substantial enough to absorb vibration and to form permanent, rigid support for the base plate. A concrete foundation on a solid base should be satisfactory.

3. It is recommended that the services of a manufacturer's erecting engineer be employed to install and start any pump equipment that is of appreciable value or precision.

4. Horizontal centrifugal pumps driven by an independent driver through a flexible coupling, with pump and driver mounted on a common base plate, should be realigned after the complete unit has been leveled on the foundation and realigned again after the grout has set and foundation bolts have been tightened. Check alignment after the unit is piped and recheck periodically.

5. It is desirable to support and restrain both the suction and discharge pipes near the pump to avoid applying forces and moments to the pump casing. Both a check valve and a stop valve should be installed in the discharge line. The check valve, placed between the pump and stop valve, protects the pump from reverse flow and excessive backpressure. Keep suction piping free of air leaks. A horizontal suction line should have a gradual rise to the pump. Any high point in the suction line will become filled with air and prevent proper pump operation.

6. To prevent the pump from freezing during cold weather when the pump is not in operation, the pump casing on dry pit applications should be drained by removing the bottom drain plug. For vertical wet pit pumps, removal of the unit may be required.

The pump manufacturer's instruction book must be read thoroughly before the pump is installed and placed in service.

Fiberglass pressure vessel installation. The manufacturer's instructions for installing and operating a fiberglass pressure vessel must be carefully studied and followed for product-specific information. The following guidelines provide general information:

1. Position each vessel on its mounting frame so that it is centered between heads. Drain holes must always face downward and be free of obstruction.
2. The mounting design must allow for vessel expansion, both axially and radially. Mount the vessel using the manufacturer's supplied or recommended straps and saddles. Do not use a U-bolt to hold down the vessel. Straps should be tightened to hold the vessel while still allowing for vessel expansion. Use flexible piping connections at both ends to permit decoupling of the header from the vessel. Typical connections are J-bends or flexible hoses. Grooved-end coupling connections are preferred for J-bends.
3. Feed, concentrate, and permeate headers should be supported independently. Follow the manufacturer's recommendations for the maximum allowable weight of branch piping.
4. Do not support piping or other components on the pressure vessels without first consulting the vessel manufacturer. Vessel ports must not be used to support any piping or other components.
5. End closures, also known as end cap assemblies, furnished by the different vessel manufacturers may differ in design. Follow the vessel manufacturer's instructions for building the end closures. It is recommended that the installer use only the closure assembly tools supplied by the vessel manufacturer. During assembly and disassembly of the end closures, examine all parts for any damage that could affect structural strength or sealing properties. Inspect all components for any condition that may have promoted corrosion.
6. Be careful not to scratch the inside wall of the pressure vessel, especially in the O-ring sealing area inside the retaining ring groove near each end.
7. Before using a forklift to handle the vessel, pad the forks to lessen the chance of damaging the shell.

Element loading guidelines. This section addresses general loading procedures for spiral-wound membrane elements. Before the membrane elements are loaded, the system should be hydraulically tested with an orifice plate on the permeate line. The manufacturer's information should always be consulted for detailed instructions. It is best to have two or three people for loading. Gloves should be worn during the handling of new elements to protect handlers from the biocide solution and occasional fiberglass strands. Membrane elements are shipped in sealed polyethylene bags containing a pH-adjusted biocide solution, typically a 0.5 to 1 percent solution of sodium metabisulfite or 0.2 percent solution of glutaraldehyde. For freeze protection, manufacturers may add glycerine to the biocide solution.

Appropriate gloves, safety shoes, and safety glasses should be worn for all operations discussed in this section. Figure 3-14 shows two workers manually loading an element into a pressure vessel. Figure 3-15 illustrates a machine-assisted loading of a membrane element. Elements in each pressure vessel should be loaded as follows:

1. Remove the elements from their plastic bags and stand vertically with the flow arrow pointing down.
2. Install the brine seals on each element on the upstream end, making certain the open groove of the brine seal is toward the upstream (feed) end of the element. Check that each element has a brine seal on the end opposite the direction of the arrow.

Figure 3-14 Manual loading of a membrane element into a pressure vessel

Figure 3-15 Machine-assisted loading of a membrane element into a pressure vessel

3. Lubricate the brine seals and interconnector O-rings with a glycerine and water solution or any other lubricant recommended by the membrane manufacturer. When installing O-rings, do not roll them into position.

4. Open and remove the pressure vessel end caps according to the pressure vessel manufacturer's instructions.

5. Inspect the interior of the pressure vessel for any surface damage. The interior of the pressure vessel should be perfectly smooth, without nicks or protrusions. Spray clean water through the open pressure vessel to remove any dust or other foreign matter. If additional cleaning is required, contact the vessel manufacturer for the recommended procedure. Lubricate the inside of the entrance of the pressure vessel with glycerine.

6. Carefully pick up the first element and slide it into the pressure vessel, making sure that the brine seal is properly seated in the groove on the feedwater end of the element. The elements should be loaded in the same direction as the feedwater flow. Allow the element to protrude from the pressure vessel by approximately 8 to 10 in. Then install the first interconnector with O-rings into the first element with a pushing and twisting motion until it is fully seated.

7. For safety and efficiency, two persons should load the elements. The lead person places one hand firmly on the side of the first element to properly position it and guides the next element into position. With a slight twisting motion, one end of the interconnector is gently inserted into the permeate tube of the element. If external interconnectors are used, the interconnector will fit over the permeate tube of the element. Slowly push the element into the pressure vessel, leaving 6 to 8 in. protruding from the pressure vessel.

8. Repeat the preceding steps until the required number of elements have been coupled and inserted.

9. Although most manufacturers use similar interconnectors for elements, end adapters could vary depending on the vessel manufacturer. Follow the vessel manufacturer's assembly instructions to make the final connection between the element and the end cap of the pressure vessel.

10. Follow the preceding instructions for the end cap assembly.

11. Repeat the preceding steps for the remaining pressure vessels.

Element removal guidelines. It is recommended that two individuals (not just one) unload membrane elements from the pressure vessels. The general procedure outlined here, as well as all instructions provided by the system manufacturer, should be followed. After the system has been shut down, it is important to make sure the pressure from the system is completely released. All pressure gauges should read zero pressure before any pressure vessel is opened.

1. Disconnect the hard plumbing at each end of the pressure vessel. Mark all removed parts for reassembly at the same location.

2. Remove end cap assemblies according to the vessel manufacturer's instructions.

3. Push the elements from the pressure vessel in the same direction in which feedwater flows. Push the elements out one element at a time. Support

each element as it is being pushed from the vessel until the element is free of the pressure vessel.

Media filter installation. The following steps describe the initial installation procedure for media pressure filters:

1. Set the units in place, ensuring they are properly leveled.
2. Connect all face piping, if not factory assembled, and align according to drawings furnished by the manufacturer or system supplier.
3. Anchor all units to the foundation to ensure stability during operation.
4. If internal filter components (filter support, backwash troughs, air and/or water scour, etc.) are not installed, install them at this time, carefully following manufacturer's instructions and drawings. Make sure all internal filter components are properly supported.
5. Load filter media by following the manufacturer's instructions and as outlined in this step. Filter media are shipped in bags or marked containers that can weigh 50 to 150 lb depending on the kind of media used. Erect a temporary scaffold next to the vessel porthole, and use a forklift or crane to lift the bags to a platform where they can conveniently be opened and the media lowered into the tanks. A factory-trained representative should be consulted for any questions that may arise. Media should be stored unopened in a dry place until just before they are loaded in the tank. A dust mask should be worn during media loading.
    a. Remove the cover on the upper tank porthole by loosening the bolts that secure the cover.
    b. Inspect internal components visually to ensure that no damage occurred during shipping. Verify the equipment with drawings furnished by the manufacturer.
    c. Gently lower (do not pour) the filter media and level.
    d. Carefully add water as the media are installed to level each layer and eliminate air pockets.
    e. Remove and account for all materials and tools used in loading before closing the tank with the porthole cover.
    f. Verify that the gaskets are well seated and properly aligned so that the porthole cover is secure.
6. Connect service piping to the unit, supporting and securing all piping to prevent excessive vibrations and strains.
7. Connect drain piping to the unit.

Reverse osmosis and nanofiltration unit installation. Reverse osmosis and NF units may arrive largely preassembled or may require some field assembly. The following installation guidelines provide an idea of the range of preassembly available.

Skid-mounted units. Skid-mounted units are delivered preassembled. Typically a concrete slab is provided to support the skid. The skid is placed on top of this slab and anchored to it where the manufacturer has provided bolt holes. The only field piping connections required are for the feedwater, concentrate, and permeate. The only electrical connection needing installation should be the power supply to the skid-mounted panel containing the motor starter and control circuitry.

Field-erected units. Because of their physical dimensions, larger RO and NF units are field erected. The system manufacturer furnishes a prefabricated steel or fiberglass structure for assembly on the jobsite in order to support the pressure vessels. Either type of structure requires a concrete pad with anchor bolts provided as indicated by the manufacturer-supplied plot plan. The erector determines whether or not grouping of the support frame is required.

The first step in the assembly is placement of the vertical frame sections. Laser technology may be employed to arrive at a properly aligned assembly. If grouping is required, a minimum allowance of 1 in. should be made. If the frame rests directly on top of the concrete, a sealing compound should be applied between the concrete and the frame to keep water out.

The next step involves installing sway supports and straps to tie the different sections together. All fasteners used in the assembly should be stainless steel, with flat washers under the bolts and the nuts. Any repairs to steel coatings or FRP should be made at this point because access to the frame is nearly impossible once the pressure vessels are in place.

Pressure vessels should be installed according to the manufacturer's instructions. To ease installation, the process can start with the farthest vertical row first. If necessary, the vessels should be turned so that any identification code is visible. The saddles and straps provided by the vessel manufacturer should be used to secure the vessels to the frame. Installation continues with the next vertical row until all vessels have been installed. After all vessels have been installed, the feed, concentrate, and permeate headers should be placed in their respective positions. Connections between these headers and the pressure vessels are established with J-bends or hoses according to the design.

The membrane elements are installed only after the system has been hydrotested and has met disinfection requirements. It is important to follow the membrane manufacturer's guidelines to install elements. A qualified representative from the membrane manufacturer's facility should be present during the installation of elements and system startup. Special precautions should be taken in case the membrane elements have no tolerance for oxidants such as chlorine. Before elements that require an oxidant-free environment are loaded, the system should be flushed with nonchlorinated water. It may be necessary to flush the system with sodium bisulfate solution.

Instrumentation and Controls

The range of controls may vary from simple instrumentation and manual controls to sophisticated controls requiring a specialist for installation and maintenance.

The basic instrumentation required to control and monitor any RO system consists of control valves and devices for measuring

- flow
- pressure
- conductivity
- pH
- liquid level
- time

Instrumentation, in its simplest form, may consist of rotameter flowmeters; Bourbon-type, direct-reading pressure measurement; and insertion-type probes for pH, conductivity, and local indications. More complex instrumentation would have 4 to 20-mA or 0-5-V direct current signal transmitters for remote monitoring and control. Larger systems include programmable logic controllers and a personal computer system for operator interface. Wiring diagrams furnished by the system supplier should be strictly followed. After installation, controls should be checked for accuracy and proper functioning by simulating operating conditions.

Instrument location is determined during the design phase of the project. The installer should make sure that the instruments are adequately secured and will not be subjected to significant vibrations and undue turbulence. Instruments should be easily accessible because most instruments require frequent calibration and repairs. Guidelines published by the Instrumentation Society of America and the Hydraulic Institute should be studied and followed to select, locate, and install the instruments.

REFERENCES

Hydraulic Institute. 1994. *Centrifugal Pumps.* Standard 1.1-1.5-1994. Parsippany, N.J. Hydraulics Institute.

Catalog. 1998. The Institute of Electrical and Electronics Engineers, Inc. Instrumentation and Measurement Society of America. Pascataway, N.J.: IEEE.

Chapter 4

Operations and Maintenance

Reverse osmosis (RO) and nanofiltration (NF) plants require specialized operations and maintenance (O&M) procedures. This chapter presents guidelines for most common procedures, but these guidelines should not be viewed as a substitute for specific manufacturer's instructions. Safety is also discussed in this chapter.

Figure 4-1 shows an operator collecting data from a local instrument panel. Proper system monitoring and maintenance will maximize membrane life and minimize system downtime. The reference system shown in Figure 4-2 will be used in the following discussion. Individual RO and NF system parameters will vary from the reference system, but the guidelines given here apply to all systems. Only O&M procedures unique to membranes are discussed. Lubrication of pumps and motors, routine data collection, and other procedures common to all water treatment systems are omitted.

The most common problem with RO and NF membrane systems is fouling by suspended particles in the water (Figure 4-3). Fouling is a reduction of water flow through a given area of membrane at a given temperature, salt concentration, and pressure due to increased resistance to flow caused by particles on or embedded in the membrane and in the feed channel spaces. These particles may be biological in origin (bacteria) or nonbiological (colloids, silts, and clays). The second most common problem is scaling. Scaling is caused by a concentration of an inorganic salt in excess of its saturation point, which creates salt precipitation and deposition on or in the membrane and subsequently in the feed channel spacers. Scaling and fouling reduce the amount of permeate flow, usually adversely affect permeate quality, and increase the pressure drop across the pressure vessel.

Figure 4-1 Data collection (City of Dunedin, Fla.)

PROCESS MONITORING

System monitoring is essential to prevent fouling and scaling and to identify a problem early when it may easily be reversed. Table 4-1 lists typical monitoring parameters, tests, frequency of testing, and limiting conditions. Figure 4-4 provides an example of a typical operating log.

Pretreatment

Treating source water before it reaches the membrane system protects the unit from dissolved and suspended particles that can foul, scale, and physically or chemically destroy the membrane elements. The fouling potential of suspended substances is indicated by measurements of turbidity and silt density. The scaling potential of dissolved substances is indicated by measurements of conductivity and the saturation levels of specific scaling compounds.

Turbidity. The turbidity test measures the light-scattering ability of particles in water. As Figure 4-5 indicates, a turbidity meter passes a light beam through water. If particles are in the water, light is scattered. The amount of light scattered is measured by the instrument and indicated in nephelometric turbidity units (ntu).

Reverse osmosis and NF manufacturers generally require source water to have a turbidity value of less than 1 ntu. A continuous, on-stream turbidity measurement of surface water is suggested because of the highly variable suspended solids loading of such water. Measuring downstream of a clarifier or media filter is recommended to monitor the effectiveness of particle removal and to ensure a turbidity less than 1 ntu in the membrane feedwater. Turbidity is frequently not measured on groundwater or beach well sources because of their normally low particulate levels and low variability.

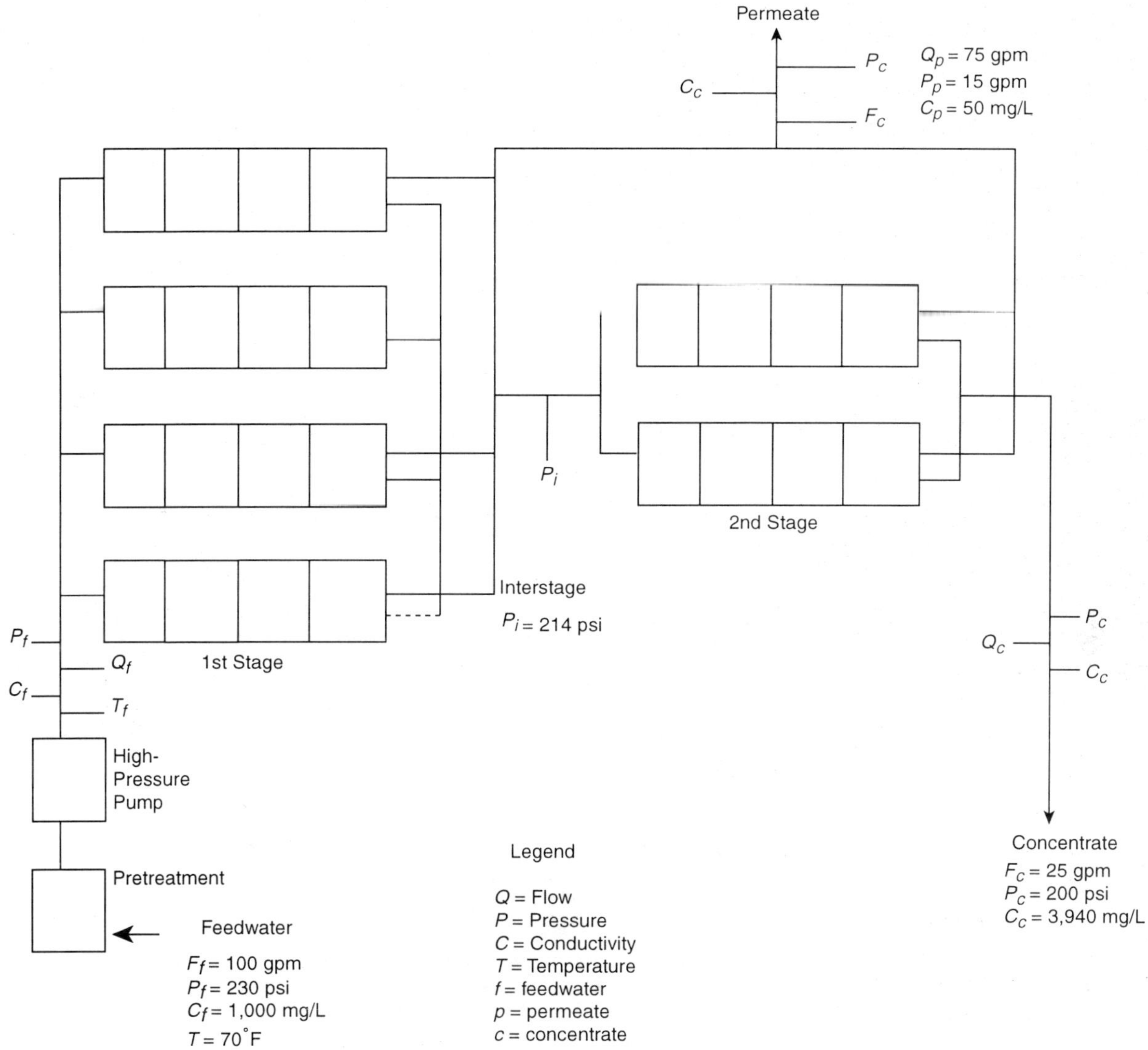

Feedwater: Groundwater or surface water, source or processed

Pretreatment: May include any or all of the following:

- pH control (generally acid addition)
- scale inhibitor addition (generally polyacrylates)
- carbon filtration (organics and chlorine removal)
- particle filtration
  - media filtration (common for surface feedwater)
  - cartridge filtration (common for all)
- microbial control
  - chlorine addition (cellulosic membranes)
  - chlorination/dechlorination (noncellulosic membranes)
  - ultraviolet light (occasionally)
  - sanitation during cleaning (common for all)
  - dechlorination (common for noncellulosic membrane)
  - carbon filtration
  - sodium bisulfite addition

High-pressure pump: Centrifugal (common) or positive displacement. May have variable speed motor drive.

Pressure vessels: Contain RO or NF membrane elements connected in series (4 shown, 1–6 common)

Permeate: Purified water (product) that has passed through the RO or NF membrane

Concentrate: Feedwater, of which dissolved and suspended constituents have been concentrated by the removal of pure water (permeate). Generally waste stream requiring disposal.

Figure 4-2 Reverse osmosis or nanofiltration process flow diagram

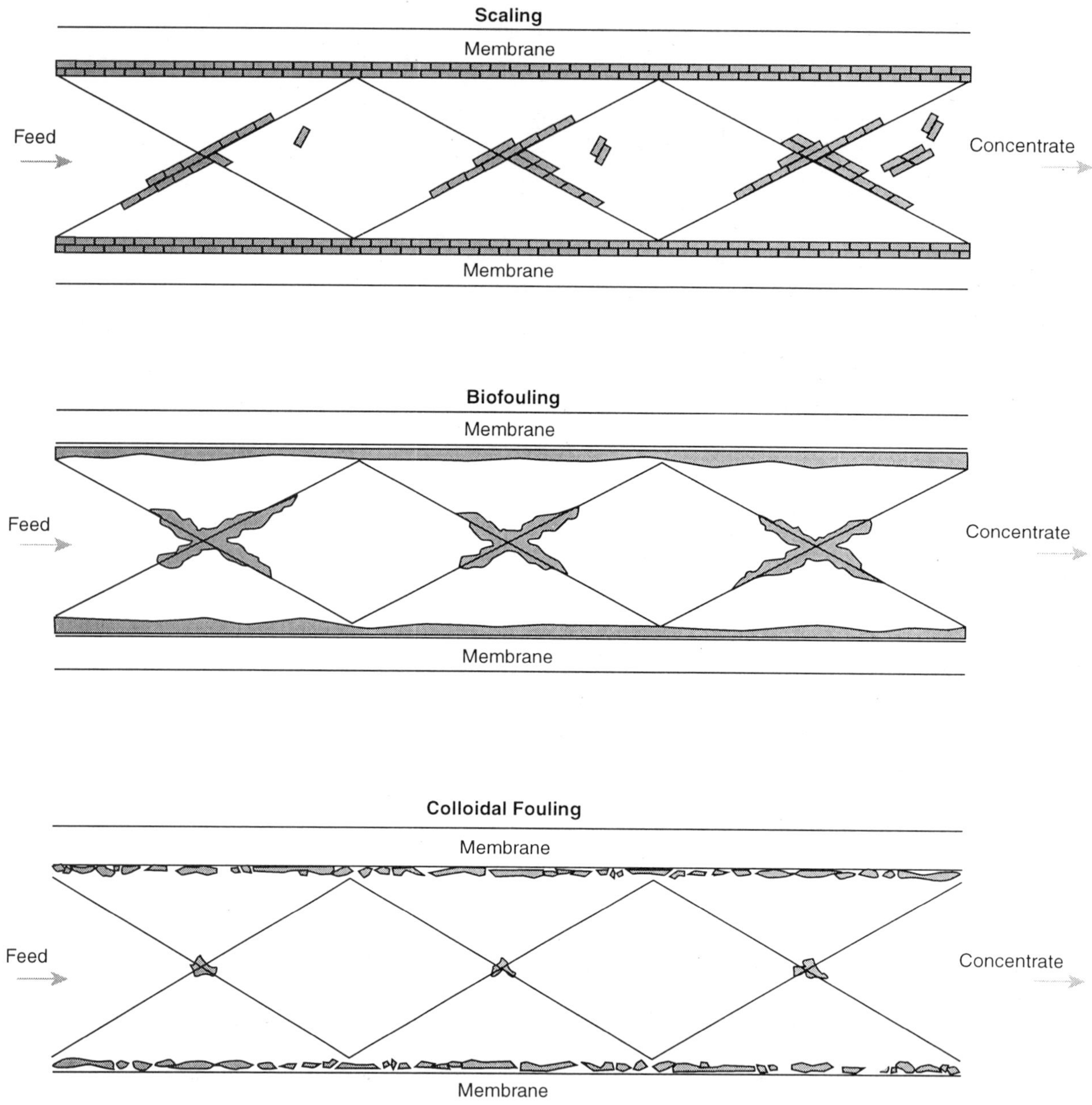

Source: Copyright 1993, David H. Paul, Inc.

Figure 4-3 Scaling and fouling

Silt density index. The silt density index (SDI) test is discussed in the pretreatment section in chapter 2. The SDI is a measure of the fouling potential of particles in water. Since fouling is the primary potential problem in RO and NF systems, this is an important measurement. Pretreatment systems should be designed, maintained, and operated to provide the lowest SDI for system feedwater. During system upsets, SDI should be measured frequently to ensure low-SDI water. If possible, the membrane unit should not be operated with a feedwater SDI higher than 5 for spiral-wound membranes and 3 for hollow fiber membranes. In any case, the manufacturer's recommendation concerning SDI levels should be followed.

Table 4-1 Monitoring parameters for reverse osmosis systems

Parameter	Measurement	Frequency	Limits	Comments
Pretreatment monitoring:				
Feedwater particulates (monitoring fouling potential)	Turbidity, ntu	Continuously, on stream	<1 ntu*	Surface water
		1x/mo. and during system upsets	<1 ntu	Groundwater
	Silt density index, SDI	3x/day	<5 ntu†	Surface water 24-hour operation
		1x/day	<5 ntu	Surface water 8-hour operation
		1–4x/mo. and during system upsets	<5 ntu	Groundwater
			$CaCO_3$ (LSI) ≤ + 1.5	See text
		1x/week	$BaSO_4$ ≤ 60 × ksp	See text
			$CaSO_4$ ≤ 2 × ksp	See text
			$SrSO_4$ ≤ 8 × ksp	See text
Feedwater dissolved substances (monitoring scaling potential)	Alkalinity, barium, calcium, pH, strontium sulfate	Continuously, on stream	None	See text
		4–12x/year	Less than scaling level in concentrate*	Surface water
	Conductivity, μmhos/cm or μS/cm	1x/year		
	Saturation (scaling level), percent saturation		Less than scaling level in concentrate	Groundwater
Particle counts and size distribution	Particle counter	1x/day	Based on feedwater levels	Determines relative changes in particle density and distribution
Membrane unit monitoring:				
Flow	Feedwater, permeate, concentrate	Continuously, on stream	Design specification	See text
	Percent recovery	Continuously, on stream	Design specification	See text
	Normalized permeate flow	Daily, graphed weekly	Design decline	10–15 percent drop requires cleaning
	Pressure drop	Daily, graphed weekly	Design specifications	See text
Salt concentration (conductivity)	Feedwater, permeate, concentrate	Continuously, on stream	Design permeate quality	See text
	Percent salt rejection	Daily, graphed weekly	Design rejection	See text
	Permeate from each pressure vessel	Monthly	Trend increases	See text
Temperature	Feedwater	Continuously, on stream	Design specification	Surface water
		Continuously, on stream	Design specification	Groundwater
Pressure drop, normalized permeate flow (see above)	Feedwater, permeate, concentrate, interstage	Continuously, on stream	Design specifications	See text
Biological monitoring:				
Chlorine dosage	Feedwater	1–3x/day	Generally 0.5–1.0 mg/L	Cellulosic membranes only
	Feedwater	1–3x/day or continuously†	Generally 0.0 mg/L	Noncellulosic membranes
Bacterial counts	Feedwater, permeate, concentrate	2–4x/year, usually	Trend increases	Systems that cannot continuously add biocide
Instrumentation calibration:				
Flows, pressures, conductivities, turbidities, other	Feedwater, permeate, concentrate	1–4x/year, usually	Scheduled preventive maintenance	See text
System balance check	Flows and conductivities	Weekly	Within 5–10 percent	See text

*Includes less than scaling level with addition of acid, or scale inhibitor, or both.

†Most noncellulose membranes require dechlorination by activated carbon or sulfite addition. An on-stream, oxidation–reduction potentiometer or chemical analyzer may be used.

	Times Per Shift ______ Train #	Design							
	Date								
	Time								
	Operating Hours								
Pressure (pig)	Feed 1st Stage								
	Feed 2nd Stage								
	Permeate								
	Concentrate								
D (pig)	Cartridge								
	1st Stage								
	2nd Stage								
Flow (gpm)	Feed								
	Permeate								
	Concentrate								
	Recovery (%)								
Conductivity (mho/cm)	Feed								
	Permeate								
	Concentrate								
TDS (mg/L)	Feed								
	Permeate								
	Concentrate								
	Salt Passage (%)								
pH	Source Water								
	Feed								
	Concentrate								
	Permeate								
Feed	C (mg/L)								
	SDI								
	Turbidity (ntu)								
	Temperature (°C)								
Acid	Level								
	Refill (1)								
	Consumption (g/m^3)								
Inhibitor	Level								
	Refill (1)								
	Consumption (g/m^3)								
Normalized	Permeate Flow (gpm)								
	Salt Passage (%)								
	Remarks								

Figure 4-4 Example of a membrane facility operating log

Clarifiers, media filters, and cartridge filters are the most common processes used to remove particulate matter. Filter backwashes, coagulant and polyelectrolyte dosages, and cartridge filter replacements should be carefully monitored and adjusted to give the lowest practical SDI value.

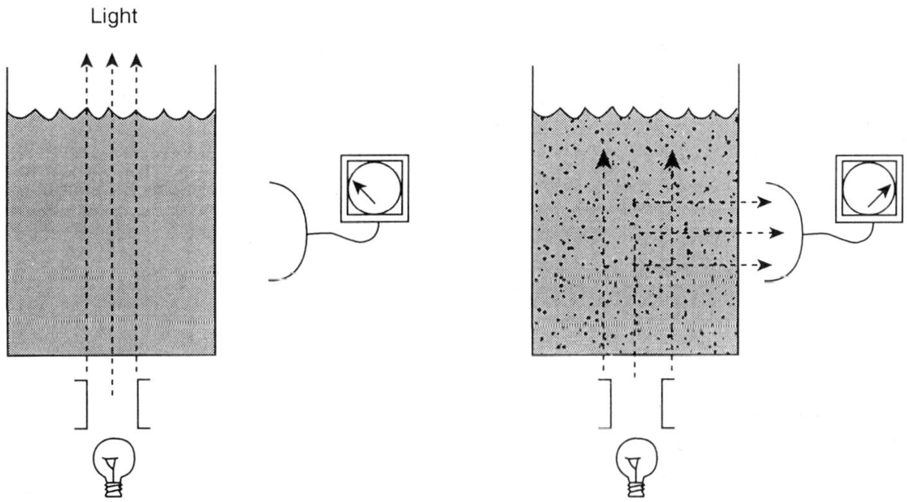

Figure 4-5 Turbidity meter passing a light beam through water

Conductivity. A conductivity meter measures the charged, dissolved substances (generally inorganic salts) in the water, not the uncharged substances, such as organics or silica. Measuring the conductivity of feedwater is necessary to track changes in the dissolved substance levels of feedwater or pretreated feedwater and to perform the monitoring calculations discussed later. Because of its relative ease of measurement, conductivity is often used as a surrogate measure of total dissolved solids (TDS). For a particular water, a constant relationship between conductivity and TDS can be established.

Saturation level. The concentration at which any scaling compound in the RO or NF feedwater reaches saturation must be determined because a value in excess of the compound's saturation point leads to scaling. The saturation level of each dissolved constituent depends on several factors, such as pH, temperature, and ionic strength (a measure of all of the dissolved substances and their charges).

An antiscalant can raise the concentration at which compounds precipitate or scale. An optimal dosage of antiscalant will inhibit scaling, but overdosing can create other problems in the system. When an antiscalant is in use, the RO or NF system should be flushed promptly on shutdown to prevent postshutdown precipitation.

Saturation levels can be predicted from existing tables and graphs, but most facilities send samples to water analysis laboratories. Laboratories that routinely deal with RO or NF systems report on the saturation levels of the common scaling compounds (calcium sulfate, calcium carbonate, barium sulfate, strontium sulfate, silica, and calcium fluoride).

Because of the variability of surface waters, feedwater from such sources should be tested as frequently as is practical for the first year or two until the seasonal variabilities of the scaling ion concentrations are known. For groundwater, testing once a year is adequate in most cases.

Membrane System

Several membrane system parameters, including flow, pressure, and temperature, must be monitored to ensure the system is working efficiently.

Flow. Every RO or NF system is designed to produce no more than a specified amount of permeate so that the salts left behind cannot concentrate in excess of their scale-forming levels. Monitoring flow rates is essential for this reason. Flowmeters should be calibrated two or four times a year according to the manufacturer's specifications. A weekly check can be made using the mass balance equation for water flow (see Eq 1-1, page 15).

Example. If measured Q_f = 100 gpm, the measured Q_p = 75 gpm, and measured Q_c = 20 gpm, then Eq 1-1 can be written as follows:

$$Q_f = Q_p + Q_c$$

100 gpm = 75 gpm + 20 gpm

100 gpm = 95 gpm (The values do not match, so flowmeter calibration is needed.)

The following equation indicates how much the RO or NF process is concentrating the feedwater:

$$CF = \frac{Q_f}{Q_c} \tag{4-1}$$

Where:

CF = concentration factor, dimensionless

Example. For flow values of Q_f = 100 gpm and Q_c = 25 gpm, the concentration factor is found as follows:

$$CF = \frac{100 \text{ gpm}}{25 \text{ gpm}} = 4.0$$

This calculation indicates that the concentration factor is 4. Since most of the dissolved substances and all of the suspended substances that were in 100 gpm of feedwater are now in 25 gpm of concentrate, the substances are four times more concentrated in the concentrate than in the feedwater.

If a decision is made to increase or decrease recovery, the concentration factor will change.

Example. If the feedwater flow is 100 gpm and the permeate flow is increased from 75 gpm to 80 gpm (i.e., an increase in recovery from 75 percent to 80 percent), the concentrate flow will correspondingly decrease from 25 gpm to 20 gpm. The new concentration factor will be

$$CF = \frac{Q_f}{Q_c} = \frac{100 \text{ gpm}}{20 \text{ gpm}} = 5$$

In this example, increasing the permeate production 5 gpm causes salts to be concentrated five times instead of four times. Note that the design value in Figure 4-2 (Q_p = 75 gpm) has been exceeded. Increasing concentration in excess of the design level can lead to scaling problems.

Pressure. Two pressures are critical to monitor: feedwater pressure (P_f) and pressure drop or differential pressure (ΔP across the membrane stages). The pressure drop from feed to concentrate is a measure of the flow resistance through the membrane system. After start-up (generally after two days of operation), the pressure drop across each stage of the system should be measured to determine the resistance to flow in the clean, new system. An increase in the transmembrane pressure drop (given the same feedwater temperature and flow rate) indicates an increased resistance to flow caused by fouling or scaling. The pressure drop is simply the feed pressure minus the concentrate pressure:

$$\Delta P = P_f - P_c \tag{4-2}$$

Where:

ΔP = pressure drop, in psi
P_f = feedwater pressure, in psi
P_c = concentrate pressure, in psi

Example. For the reference system in Figure 4-2, the pressure drop across all stages is

$$\begin{aligned}\Delta P &= P_f - P_c \\ &= 230 \text{ psig} - 200 \text{ psig} \\ &= 30 \text{ psi}\end{aligned}$$

The pressure drop across the first stage is

$$\begin{aligned}\Delta P &= P_f - P_i \\ &= 230 \text{ psig} - 214 \text{ psig} \\ &= 16 \text{ psi}\end{aligned}$$

The pressure drop across the second stage is

$$\begin{aligned}\Delta P &= P_i - P_c \\ &= 214 \text{ psig} - 200 \text{ psig} \\ &= 14 \text{ psi}\end{aligned}$$

The feedwater pressure is a measure of how much pressure it takes to produce the design quantity of permeate. If the pressure increases, given the same feedwater temperature, flow rate, and salt concentration, then fouling or scaling, or both, are requiring more pressure to be applied to produce the same amount of permeate.

Temperature. The temperature of the feedwater to the RO or NF system must be monitored to allow the necessary calculations to be performed to document trends in membrane performance, as described later in this section. As the temperature of water increases, its viscosity decreases, requiring less pressure to force a given amount of water through the membrane.

Membrane Performance

When membrane performance is calculated to be less than normal, it is likely that scaling, fouling, chemical attack, or some other membrane damage has occurred. Three performance parameters should be continually monitored: percent salt rejection (SR), normalized permeate flow, and pressure drop (ΔP). Most membrane manufacturers supply software programs for monitoring and their use is highly recommended. The equations in this section are relatively simple, but they are not as accurate as the specific equations used in the software programs.

Percent salt rejection. Each membrane has the ability to reject a certain percentage of salts found in feedwater. Most RO membranes can reject 90 to 99 percent of feedwater salts. Most NF membranes can reject 75 to 85 percent of feedwater salts, depending on water quality. One way of calculating actual SR is as follows:

$$SR = \frac{C_f - C_p}{C_f} \times 100 \tag{4-3}$$

Where:

C_f = feed salt concentration, in mg/L
C_p = permeate salt concentration, in mg/L
SR = percent salt rejection

Example. For the reference system in Figure 4-2, the feed salt concentration is 1,000 mg/L and the permeate salt concentration is 50 mg/L. Thus, the percent SR is

$$SR = \frac{1,000 \text{ mg/L} - 50 \text{ mg/L}}{1,000 \text{ mg/L}} \times 100 = 95 \text{ percent}$$

Alternatively, SR is often calculated based on the average salt concentration that is rejected. In the reference RO unit of Figure 4-2, feed salt concentration is 1,000 mg/L and concentrate salt concentration is 3,940 mg/L. The average feed/concentrate salt concentration is 2,470 mg/L [(1,000 + 3,940)/2]. The average percent SR then becomes the following:

$$SR = \frac{2{,}470 - 50}{2{,}470} \times 100 = 98 \text{ percent average}$$

Note that there are more precise (and more complex) logarithmic equations that may be used to estimate the average feed salt concentration, but those equations are not shown here.

In comparisons of membrane performances, a consistent method of calculating should be used. The percent salt passage is equal to 100 minus the SR. The reader is cautioned that a decrease in salt passage from 99 percent to 98 percent would result in roughly a doubling of salt in the permeate. If SR decreases, a scaling or fouling layer has formed on the membrane, the membrane has become damaged (via hydrolysis, chemical, or chlorine attack; or tears, cuts, or holes), or feedwater is bypassing the membrane (via O-ring leakage).

Normalized permeate flow. Four factors determine how much permeate will be produced per given area of membrane: membrane compaction, pressure, temperature, and feedwater salt concentration. Some membranes compact over time because of temperature and pressure effects. This compaction causes the permeate flow to decrease, but the membrane supplier compensates for this phenomenon when designing the RO system. As feedwater pressure increases at a constant temperature and feedwater salt concentration, more permeate is forced through the membrane. As temperature increases at constant pressure and feedwater salt concentration, more permeate is forced through the membrane. As feedwater salt concentration increases at constant temperature and pressure, less permeate is forced through the membrane as a result of the higher osmotic pressure.

When an RO or NF system is new, a certain quantity of permeate will be produced given the start-up temperature, feedwater pressure, and feedwater salt concentration. If, say, after six months of operation, less permeate is produced at the same temperature, pressure, and salt concentration (accounting for membrane compaction), then scaling, fouling, or other problems have occurred.

To allow a problem with permeate flow to be recognized, the flow can be normalized to take expected changes due to compaction into account. The current amount of normalized permeate can be compared with either the amount of normalized permeate produced by the new system (monitored at start-up) or projected values provided by the membrane manufacturer. If the normalized permeate quantity is nearly equal to start-up values, with membrane compaction accounted for, then the membrane is clean. If permeate quantity is significantly reduced (10 to 15 percent less), a scaling or fouling layer is probably inhibiting permeate production.

To calculate the cleanliness of the membrane by normalizing permeate flow, either of the following equations may be used:

$$\mathrm{NPF} = \frac{\mathrm{NDP_{su}}}{\mathrm{NDP_{today}}} \times \mathrm{TCF}_{77} \times Q_p \times \frac{1}{\mathrm{MC}} \tag{4-4}$$

$$\mathrm{NPF} = \frac{\mathrm{NDP_{su}}}{\mathrm{NDP_{today}}} \times \frac{\mathrm{TCF_{su}}}{\mathrm{TCV_{today}}} \times Q_p \times \frac{1}{\mathrm{MC}} \tag{4-5}$$

Where:

$\mathrm{NDP_{su}}$	=	net driving pressure at start-up, in psi
$\mathrm{NDP_{today}}$	=	net driving pressure today, in psi
NPF	=	normalized permeate flow, in gpm
$\mathrm{TCF_{77}}$	=	temperature correction factor to normalize to operation at 77°F (25°C)
$\mathrm{TCF_{su}}$	=	temperature correction factor at start-up
$\mathrm{TCV_{today}}$	=	temperature correction value for today
Q_p	=	gauge permeate flow (actual flow of permeate), in gpm
MC	=	membrane compaction factor, supplied by manufacturer, as affected by time, pressure, and temperature

NOTE: TCF and TCV values are provided by the membrane manufacturer in the form of tables or equations.

Normalized permeate flow *must* be monitored; it documents the status of the membrane system and determines when the system needs to be chemically cleaned.

Net driving pressure. Net driving pressure is the pressure available to force water through the membrane; it is equal to the pressure on the feedwater side of the membrane minus the backpressures (osmotic and permeate backpressure). Osmotic pressure is a function of the types and amounts of salts in the water. A rule of thumb is that for every 100 mg/L of TDS, there is about 1 psi of osmotic pressure. Therefore, multiplying the salt concentration by 0.01 provides an estimate of the osmotic pressure. The average net applied pressure may be expressed as

$$\overline{\mathrm{NDP}} = \bar{P}_f - \bar{P}_o - \bar{P}_p \tag{4-6}$$

Where:

$\overline{\mathrm{NDP}}$	=	average net driving pressure, in psi
$\bar{P}_f$	=	average feedwater side pressure, in psi, $= \dfrac{P_f + P_o}{2}$
$\bar{P}_o$	=	average osmotic backpressure, in psi, $\approx \left(\dfrac{C_f - C_c}{2}\right) - C_p \times 0.01$
$\bar{P}_p$	=	permeate backpressure, in psi

NOTE: In calculating $\bar{P}_o$, the osmotic pressure of the permeate stream is relatively low in comparison to the feed and concentrate water streams and is often ignored.

Example. For the reference setup in Figure 4-2, $\overline{\text{NDP}}$ may be determined as follows:

$$\bar{P}_f = \frac{P_f + P_c}{2} = \frac{230 + 200}{2} = 215 \text{ psig}$$

$$\bar{P}_o \approx \frac{C_f + C_c}{2} \times 0.01 = \frac{1{,}000 + 3{,}940}{2} \times 0.01 = 25 \text{ psig}$$

$$\bar{P}_p = 15 \text{ psig}$$

$$\begin{aligned} \overline{\text{NDP}} &= \bar{P}_f - \bar{P}_o - \bar{P}_p \\ &= 215 \text{ psi} - 25 \text{ psi} - 15 \text{ psi} \\ &= 175 \text{ psi} \end{aligned}$$

Temperature correction factors. Membrane manufacturers supply temperature correction factors (TCFs). Temperature correction factors allow the amount of permeate forced through a given area of membrane (at a given feed pressure and salt concentration) to be referenced to the amount that would be produced at 77°F (25°C). Two types of TCF data may be supplied. The first type, shown as TCF_{77} in Eq 4-4, indicates how much permeate would be produced at 77°F (25°C) compared to what is being produced at the current temperature. For example, for the TCF_{77} values provided in Table 4-2, an RO system producing 75 gpm of permeate at 60°F would produce 97.5 gpm of permeate at 77°F, given the same net driving pressure.

The second type of TCF, shown as TCF_{su} in Eq 4-5, provides the percentage of flow at 77°F that would be produced at other temperatures. Table 4-3 gives an example of how such values might be listed.

Table 4-2 Sample TCF_{77} values for use in normalizing permeate flow to 77°F

Temperature, *°F*	Actual Permeate Flow, *gpm*		TCF_{77}, Value Provided by Manufacturer		Normalized Permeate Flow,* *gpm*
90	75	×	0.8	=	60
80	75	×	0.95	=	71.3
77	75	×	1.0	=	75
70	75	×	1.1	=	82.5
60	75	×	1.3	=	97.5

*Given same net driving pressure.

Table 4-3 Sample TCF_{su} (start-up TCF) values for use in expressing current permeate flow as a percentage of the flow at 77°F

Temperature, *°F*	TCF_{su}	Percentage of flow at 77°F*
90	1.25	125
80	1.05	105
77	1.0	100
70	0.9	90
60	0.75	75

*Given same net driving pressure.

Example. For the reference system, assume a temperature of 60°F, and MC = 1. Equation 4-4 compares NDP at start-up to current (today) and yields

$$\text{Start-up NPF} = \frac{NDP_{su}}{NDP_{today}} \times TCF_{77} \times \frac{Q}{F_p} \times \frac{1}{MC}$$

$$= \frac{175 \text{ psig}}{175 \text{ psig}} \times 1.3 \times 75 \text{ gpm} \times \frac{1}{1}$$

$$= 97.5 \text{ gpm}$$

For the same situation, Eq 4-5 yields

$$\text{Start-up NPF} = \frac{NDP_{su}}{NDP_{today}} \times \frac{TCF_{su}}{TCF_{today}} \times \frac{Q}{F_p} \times \frac{1}{MC}$$

$$= \frac{175 \text{ psig}}{175 \text{ psig}} \times \frac{0.75}{0.75} \times 75 \times \frac{1}{1}$$

$$= 75 \text{ gpm}$$

An NPF calculation should be obtained daily and graphed weekly. Each subsequent NPF measurement should be compared to the NPF at start-up. When the NPF drops 10–15 percent relative to initial start-up values, the system should be chemically cleaned.

Pressure drop. The pressure will decrease across each membrane element and through associated piping because of the resistance to flow. The pressure drop value varies with flow rate (increased flow means increased pressure drop) and inversely with temperature (increased temperature means decreased pressure drop). An increase in the pressure drop at the same feedwater temperature and flow rate indicates a greater resistance to flow than demonstrated by new membrane elements, signifying fouling by particulate matter or scaling by dissolved substances that have precipitated. (See the pressure drop calculations described previously.)

BIOLOGICAL MONITORING

Fouling of noncellulosic RO and NF membrane systems by bacteria is a constant threat because most noncellulosic membranes do not tolerate continuous additions of an oxidizing biocide, typically chlorine, in the feedwater. Bacterial growth is minimized by operating the system at design flow rates and by maintaining proper pretreatment chemical feeds. The shear force imparted by the feedwater stream of an operating system limits bacterial growth. When a system is shut down, however, bacterial biofilm formation may increase. Within 24 to 48 hours of nonoperation, bacteria multiply and produce substances around their bodies that effectively adhere to the membrane system. In addition, seawater RO systems can be adversely affected by microorganisms, such as algae, if pretreatment systems fail to operate properly.

If the RO or NF unit must be shut down, it should be flushed with permeate or feedwater at least once a day to limit bacterial growth. Flushing or operating the system every 8 hours, if possible, is recommended. A biocide should be added to the system if the system will be shut down for two or more days.

Bacteria are usually quantified by several measures or methods, including the standard plate count, field cassette, direct count (via electron microscopy and epifluorescence), total organic carbon, Lamellas amoebocyte lysate, and, most commonly, heterotrophic plate count. See *Standard Methods* (APHA, AWWA, and WEF 1995) for analytical procedures. A significant increase in bacteria exiting RO or NF pressure vessels indicates excessive growths in the system, requiring chemical cleaning and sanitization.

Bacterial layers cause significant increases in resistance to flow. As discussed previously, when biofouling occurs, the pressure drop across the system will rise and NPF will be decreased. The need for sanitization or chemical cleaning is determined by monitoring pressure drop, NPF, and bacterial counts.

All major manufacturers of RO and NF membranes provide computer software for data normalization. The software takes into account all membrane variables to provide accurate measures of normalized performance that should be compared with the actual measured values for the system. Again, it is recommended that these programs be used for monitoring.

CHEMICAL CLEANING

If an RO or NF system becomes fouled or scaled, the fouling or scaling material has to be removed by chemical cleaning. Chemical cleaning will accomplish the following:

- dissolve and remove inorganic scales
- flush out particulate material
- break down bacterial slimes
- eliminate bacteria

Figure 4-6 illustrates a typical cleaning system for an RO or NF unit. Chemicals used for cleaning must be compatible with the membranes and the construction materials. The mix tank should be sized for a minimum of three minutes detention time. A cover, exhaust fan, heater, mixer, and temperature indicator should be provided. A 50-micron bag filter should be used to capture suspended solids and particulate foulants released during cleaning. The cleaning stream should be in the same direction as normal feed to the RO or NF system. Cleaning is facilitated by high recirculation rates, increased temperature, and minimal permeation rates.

Chemical cleaning agents fall into the following categories:

- enzymes to break down bacterial slimes
- detergents and surfactants to resuspend particulate material and dissolve organic material
- biocides to kill bacteria
- chelators to remove scale
- acids to dissolve inorganics
- caustics to dissolve organic substances and silica

Membrane manufacturers recommend certain cleaners, and membrane chemical cleaning vendors sell proprietary chemical cleaners. Only those chemicals that have been approved in writing by the membrane manufacturer should be used in the membrane system. The membrane manufacturer's guidelines must always be followed for types of cleaning chemicals, solution temperature, pH, and other parameters. Table 4-4 shows the suitable chemical cleaning methods for different types of problems.

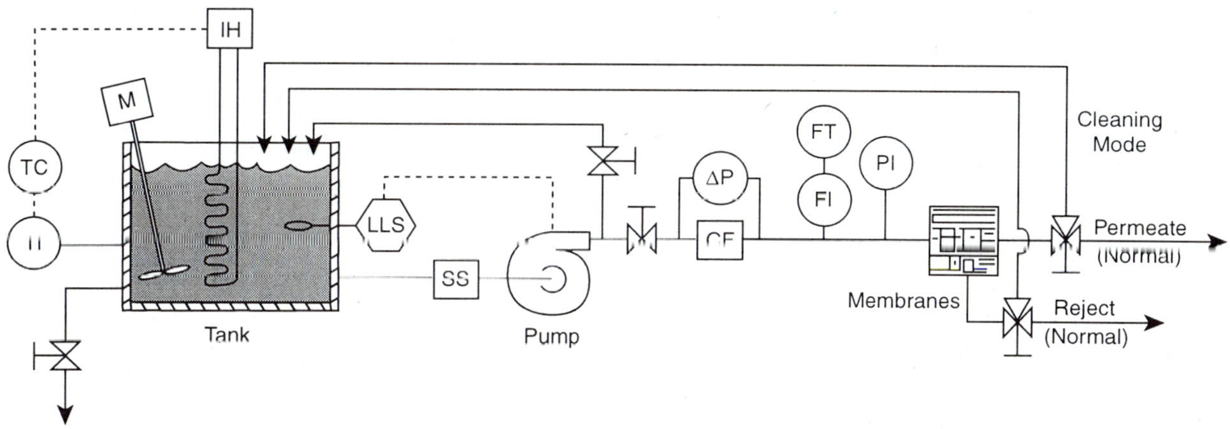

Legend

ΔP: Differential pressure gauge
CF: Cartridge filter
FI: Flow indicator
FT: Flow transmitter
IH: Immersion heater
LLS: Low-level switch
M: Motor (for mixer)
PI: Pressure indicator
SS: Silt strainer
TC: Temperature controller
TI: Temperature indicator

Figure 4-6 Cleaning system flow diagram

Table 4-4 Chemical cleaning approaches for different problems

Problem	Chemical Cleaner
Inorganic scale	Acid (low pH) Chelator
Silt or dirt	Caustic (high pH) Chelator Detergent
Biofouling	Caustic (high pH) Enzyme Detergent Chelator Biocide

Alternative procedures for chemically cleaning any membrane system are basically variations of the following minimal requirements:

1. Mix and heat the cleaning solution to desired parameters.
2. Circulate the cleaning solution through the RO or NF membranes for 30–60 minutes, discharging the first 20 percent to waste.
3. Allow the system to soak for 1–12 hours.

4. Flush the system after chemical cleaning for 60 minutes, discharging the permeate and concentrate to waste.

5. Put the system back in service.

A system should be chemically cleaned when the pressure drop across either or both stages increases 15 to 25 percent, or when the NPF drops 10 to 15 percent less than what it should be for the age of the membrane. Failure to clean at an early stage may lead to irreversible fouling or scaling. It is always easier to clean at an early stage than at a later stage. It is important to be sure that all chemicals are compatible with the membranes and associated equipment.

Each stage is generally cleaned separately. Performing multiple cleanings with the same cleaner or alternating among different cleaners is sometimes necessary to thoroughly clean the system. The system is clean when the pressure drop and the NPF have returned to the projected values.

MECHANICAL INTEGRITY

The performance of any RO or NF membrane system is determined by monitoring the quality and quantity of permeate produced by each train in the system. When one or more factors are outside projected values, the problem should be investigated and corrective action taken as soon as possible. If the problem is left unchecked, permeate water quality may deteriorate, and the loss in performance may become irreversible. Probing is a method of checking the mechanical integrity of the membrane.

Probing the Vessels

When analysis of the system indicates high salt passage, a two-step investigation is recommended to locate the source of the problem. The first step in the investigation—the system profile—measures conductivity from each vessel to identify which vessels have high conductivity. The system profile should be taken between 24 and 48 hours of system operation after start-up to establish a baseline record. To facilitate record keeping, each vessel's reading should be recorded on a diagram showing a series of circles arranged similarly to the rack assembly.

Next, the vessels that have high conductivity should be probed. Probing is a technique used to determine the relative performance of each element within a pressure vessel and to locate any elements or O-ring leaks that are the source of the problem. Conductivity measurements are taken by probing the product water tube at different locations with a plastic or metal tube. To accomplish this, a plastic or metal tube of 1/4-in. diameter and approximately 5 to 6 ft longer than the pressure vessel should be chosen. Before probing begins, the distances from the permeate plug to each O-ring and to the center of each element should be measured. The tubing should be marked so that when the probe is inserted to each mark, the operator knows where the probe is located within the vessel (Figure 4-7).

The permeate plug should be removed from the end cap (at the opposite end from the permeate manifold), and the tubing should be inserted through the length of the product water tube (Figure 4-8). To ensure that the permeate water from neighboring pressure vessels does not flow into the vessel being probed, a seal of some type is required around the tubing to force the water into the permeate manifold. A rag wrapped around the probe tubing and held against the opening in the permeate port works well as a seal, provided the permeate backpressure is not too high.

To begin the probing process, the probe tubing should be pushed through to the opposite end of the pressure vessel with short oscillating motions as required to pass over the interconnectors. The end point of the probe should be moved to the first

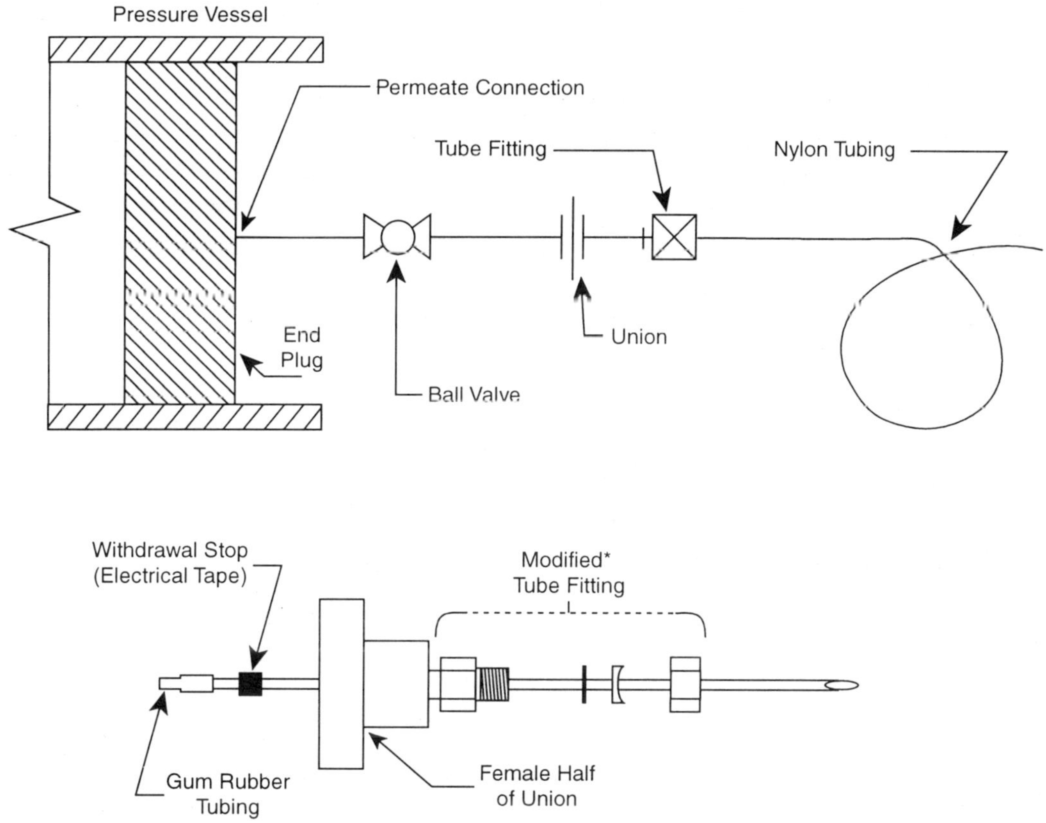

*Tube fitting modified by extending the 1/4-in. bore through the body and discarding the gripper ring.

Figure 4-7 Permeate probing apparatus for a spiral-wound membrane

sample point. A portion of the combined permeate from this area of the pressure vessel will pass into the tubing, allowing the operator to collect a sample and measure its conductivity. This location should be held until the conductivity remains constant. This usually takes a minute or so. The conductivity reading should be recorded at the appropriate position (Figure 4-9). The tubing should then be pulled out to the next mark. Once the conductivity reading has stabilized, the second reading should be recorded. The procedure should be repeated until the vessel probe is complete.

Interpreting Probe Data

As feedwater flows through a pressure vessel from one element to the next element, it becomes more concentrated. This means that permeate water from the upstream (feedwater side) element will have a lower conductivity reading than the downstream (concentrate side) element. Typically, a six-element pressure vessel will operate at about 50 percent recovery, and the concentrate exiting the vessel will have approximately twice the conductivity of the feedwater entering the vessel. The permeate conductivity from the downstream element can be expected to be approximately twice that from the upstream element. As the probe position is moved from the upstream element to the downstream element, the conductivity readings should increase uniformly.

An abrupt increase in conductivity likely indicates a leak or poorly performing element. Feedwater is sealed from the permeate stream by O-rings. If an O-ring

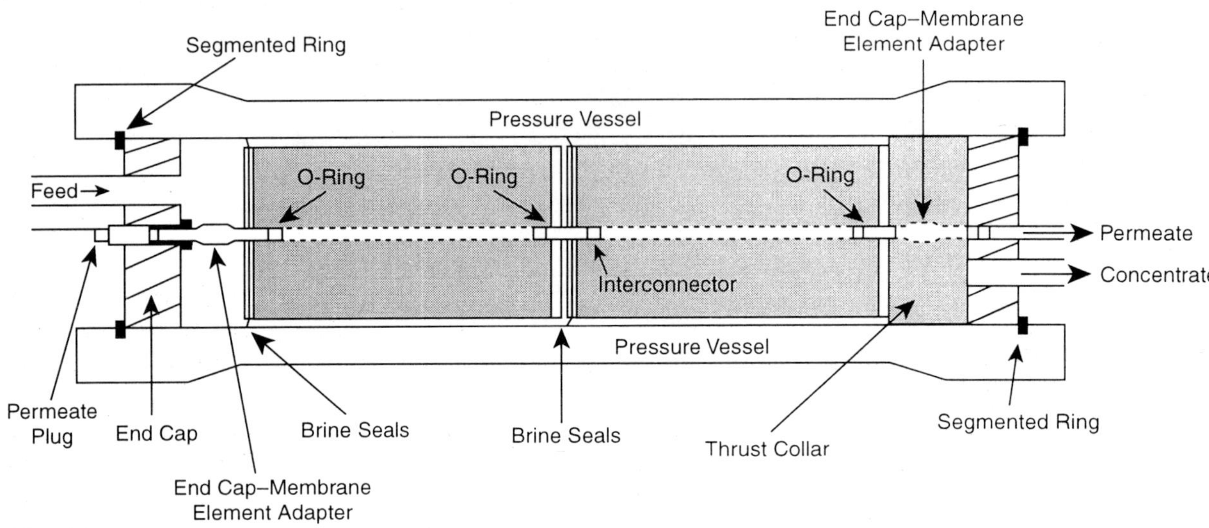

Figure 4-8 Probing a spiral-wound membrane

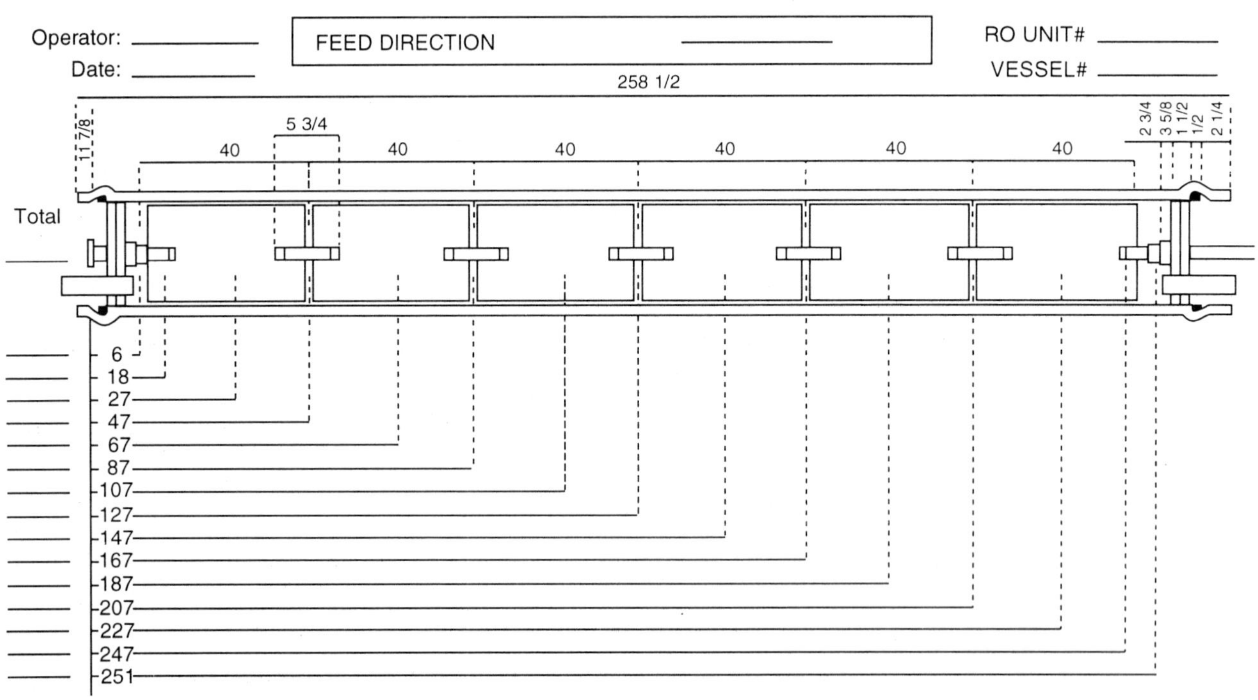

Figure 4-9 Sample pressure vessel probing sheet

becomes out of round or unseats, feedwater can flow into the permeate. Occasionally membrane elements are defective, become torn, or are chemically attacked, in which case feedwater or concentrate water leaks into the permeate stream.

Interpretation of the probe data collected by this technique is sometimes difficult because of the effects of dilution and the inherent variability or rejection characteristics between individual elements. However, the procedure can be very useful in detecting O-ring leaks and problems with the elements.

INSTRUMENTATION CALIBRATION

The performance of membrane systems is determined by monitoring pressure drop, normalized permeate flow, percent SR, and other parameters. These measurements are all performed by on-stream and bench-top instruments that must be calibrated for accuracy.

Calibration should be performed on a preventive maintenance schedule. Most instruments need to be calibrated routinely, usually two to four times per year. Calibration should initially be conducted on the schedule recommended by the instrument manufacturer, although instruments may require more frequent calibration.

To check for the accuracy of flow and conductivity instruments, a mass balance calculation for solute flux may be performed, as originally given in Eq 1-2 on page 15.

$$Q_f C_f = Q_p C_p + Q_c C_c$$

If the two sides of the equation do not match once the calculation has been performed, the instruments are in need of calibration.

SAFETY

Safety is a *fundamental responsibility* for all personnel in a water treatment plant. Industry goals include zero accidents and total compliance with all regulatory statutes. To achieve these goals, water treatment personnel involved in membrane processes need the following training and knowledge:

- a hazard review of all operating steps with appropriate follow-up audits
- operating instructions, including design criteria and prints, posted where they are easily accessible to all personnel
- training for operations, maintenance, and supervisory staff on an ongoing basis
- appropriate tools and clothing
- a thorough understanding of and compliance with Occupational Safety and Health Administration (OSHA), US Environmental Protection Agency, and other relevant types of regulations
- a facility evacuation plan, with adequate fire protection and fire-fighting equipment

The following sections address major safety considerations for an RO or NF plant. The information presented is not a complete listing but is intended to assist in efforts to start up and operate a water treatment plant safely.

General Safety Precautions

General safety precautions for RO and NF plants are no different from those for conventional treatment plants. All operators should know and follow operating instructions, OSHA requirements, and environmental regulations. Proper protective equipment (e.g., gloves, safety glasses, safety shoes, hearing protection, protective clothing, and face shields) should be worn as is appropriate.

Good housekeeping standards should be maintained. Floors should be clean and dry with excess material stored on shelves and away from operating equipment. Oily rags and waste paper should be properly discarded.

Fire exits should never be locked or blocked. Safety equipment, fire extinguishers, and hoses should be available and always in good working order. An evacuation plan should be prepared and personnel should practice it. The fire-fighting team must be adequately trained and their expertise routinely updated. All operating equipment safety interlocks and pressure relief devices should be checked. All rotating equipment should be shielded.

Materials heavier than the safe-weight standard established by the employer should not be lifted by hand. All work should be within individual ergonomic limitations. Eyewash stations and safety showers should be maintained and tested. Floor markings should clearly define safety areas as well as slipping and tripping hazards. Plant management should institute routine training programs to ensure that all personnel are thoroughly familiar with the operations, maintenance, environmental, and safety requirements for the facility.

Membrane Pressure Vessel Safety

Because RO and NF plants are pressure driven, the following safety standards relating to all pressure-driven water treatment processes need to be emphasized:

- Do not overpressurize equipment, operate at too high a temperature, or modify a pressure vessel without the expressed written approval of the manufacturer. Label equipment with these limitations.
- Follow in detail the manufacturer's written instructions for O&M of the pressure vessel. Do not assume that one manufacturer's instructions are necessarily applicable to all devices.
- All pressure vessels must be suitably anchored and restrained. Where possible, avoid standing in front of pressurized equipment.
- Use maximum care in handling, installing, dismantling, and maintaining pressure vessels. Be sure devices are depressurized before working on them.
- Install proper relief and shutdown protection devices. Regularly inspect these devices and vessels for equipment integrity.
- Minimize equipment and piping vibrations, and avoid water hammer.

Handling Chemicals

A number of chemicals are handled in a membrane plant. They must be used safely to avoid injury.

Material safety data sheets should be supplied for each chemical in the plant. The manufacturer's instructions should be read and followed and all containers and piping labeled. Unauthorized chemicals should never be used. Chemical certification from the equipment and membrane suppliers should be obtained if there are any doubts. Equipment must be properly vented, decontaminated, and flushed. All steps should be performed conforming to safety and environmental regulations. All effluent streams should be monitored for flow and composition.

It is important to follow safe and environmentally sound procedures for chemical disposal. If there are any doubts, suppliers should be asked for recommendations. Suitable protective equipment should be worn when handling chemicals, particularly acids and caustics. Emergency and containment procedures for fires, spills, and hazardous off-gases should be determined and practiced.

Treatment personnel may consider dedicating specific equipment for individual chemicals. Chemicals should not be mixed unless the chemistry is thoroughly known. Storage areas must be adequately vented safely away from the membrane plant.

Pumps, Piping, and Instrumentation

Equipment and associated instrumentation in an RO or NF plant include such items as filters, pumps, tanks, computers, compressors, valves, pipes, and racks. All must be given special consideration for their safe and proper functioning. The following are guidelines to achieve these goals:

- Always follow the manufacturer's recommendations for O&M. Written instructions should be readily available for all to use. Label all equipment and piping.
- Never exceed the specified pressure and temperature limits. Routinely inspect equipment for defects, corrosion, and other signs of abnormal wear.
- Anchor and restrain all equipment, and guard rotating parts.
- Never violate electrical safety and lockout practices, interlocks, or pressure relief devices. Inspect all instrumentation and electrical connections at least annually, particularly for grounding.
- Know the design criteria, functions, and limitations of all processing steps, equipment, and instrumentation.
- Provide adequate space around equipment for routine maintenance and, if necessary, for its removal.
- Minimize equipment and piping vibrations. Piping must be resistant to corrosion and properly designed for actual system pressures.

REFERENCE

APHA, AWWA, and WEF (American Public Health Association, American Water Works Association, and Water Environment Federation). 1995. *Standard Methods for the Examination of Water and Wastewater.* 19th ed. Washington, D.C.: American Public Health Association.

This page intentionally blank.

Appendix **A**

SI Equivalent Units Conversion Tables

	To Convert		
	Customary Units	to SI Units	Multiply by
Water Treatment			
Plant capacity	gpd	m^3/d	3.785×10^{-3}
		or ML/d	3.785×10^{-6}
Plant capacity	mgd	m^3/d	3.785×10^{3}
		or ML/d	3.785
Source water flow	gpm	m^3/s	6.308×10^{-5}
		or L/s	6.308×10^{-2}
Source water flow	ft^3/min	m^3/s	4.720×10^{-4}
		or L/s	4.720×10^{-1}
Source water temperature	°F	°C	(°F – 32) × 5/9
Chemical dosage	ppm	mg/L	1
Chemical feed rate	lb/d	kg/d	4.536×10^{-1}
Chemical feed rate	gal/h	L/h	3.785
		or ML/s	1.051
Gas feeder supply pressure	psi	kPa	6.9848
Gas feeder differential pressure	in. H_2O	kPa	2.488×10^{-1}
Gas feeder vacuum pressure	in. Hg	kPa	3.377
Displacement velocity	ft/s	m/s	3.048×10^{-1}
Power	hp	Watt	746
Settling rate	ft/h	m/h	3.048×10^{-1}
Weir overflow rate	gal/ft/d	L/m/s	3.449×10^{-3}
Filter head loss	ft	kPa	2.989
		or m	3.048×10^{-1}
Filtration rate	gpm/ft^2	m/h	2.444
		or $m^3/m^2/h$	2.444
Filter backwash rate	gpm/ft^2	$L/m^2/s$	6.789×10^{-1}

	To Convert		
	Customary Units	to SI Units	Multiply by
Water Distribution			
Area	mi^2	km^2	2.590
		or m^2	2.590×10^6
Area	acre	ha	4.047×10^{-1}
Head loss	ft	kPa	2.989
		or m	3.048×10^{-1}
Hydrant spacing	ft	m	3.048×10^{-1}
Hydraulic gradient	ft/1,000 ft	mm/m	1
		or m/km	1
Level gauging	ft	m	3.048×10^{-1}
Pipe cross section	$in.^2$	mm^2	6.452×10^2
Pipe diameter	in.	mm	2.540×10
Pipe flow velocity	ft/s	m/s	3.048×10^{-1}
Pipe length	ft	m	3.048×10^{-1}
Pressure	psi	kPa	6.895
Pump capacity	gpm	m^3/s	6.308×10^{-5}
		or L/s	6.308×10^{-2}
Pump capacity	ft^3/min	m^3/s	4.720×10^{-4}
		or L/s	4.720×10^{-1}
Residual chlorine	ppm	mg/L	1
Storage elevation	ft	m	3.048×10^{-1}
Storage volume	gal	m^3	3.785×10^{-3}
		or ML	3.785×10^{-6}
Storage volume	ft^3	m^3	2.832×10^{-2}
		or ML	2.832×10^{-5}
Water consumption	gal	m^3	3.785×10^{-3}
		or L	3.785
		or ML	3.785×10^{-6}
Water consumption	ft^3	m^3	2.832×10^{-2}
		or ML	2.832×10^{-5}
Metering	gph	m^3/h	3.785×10^{-3}
Metering	ft^3/h	m^3/h	2.832×10^{-2}

General Units for Water and Solute Mass Transfer Coefficients

General equation for water flux: $F_w = K_w P$

Where:

F_w = water flux
K_w = water mass transfer coefficient
P = pressure

Customary Units	SI Units
F_w : 1 gpd/ft^2 = 0.1337 ft/d	= 0.0407 m/d
K_w : 1 gpd/ft^2/psi = 0.0578/day^{-1}	= 0.0578/day^{-1}

General equation for solute flux: $F_s = K_s C$

Where:

F_s = solute flux
K_s = mass transfer coefficient
C = solute concentration

Customary Units	SI Units
F_s : 1 lb/ft^2/d = 0.016 ft/d	= 4.893 kg/m^2/d
K_s : 1 lb – mg/ft^2/d/L = 16.031 ft/d	= 4,886 m/d

Appendix **B**

Membrane Manufacturers for Reverse Osmosis and Nanofiltration Systems

Reverse Osmosis

Desalination Systems, Inc.	760 Shadowridge Dr. Vista, CA 92083	Phone: (619) 598-1800 Toll Free: (800) 42-DESAL FAX: (619) 598-3335
The Dow Chemical Company (Filmtec)	100 Larkin Center Midland, MI 48674	Toll Free: (800) 447-4369 FAX: (517) 638-9550
DuPont Company	DuPont Nylon "Permasep" Products P.O. Box 6101 Newark, DE 19714	Phone: (302) 451-9681 FAX: (302) 451-9686
Fluid Systems Corporation Anglian Water Group	10054 Old Grove Rd. San Diego, CA 92131	Phone: (619) 695-3840 Toll Free: (800) 525-4369 FAX: (619) 695-2176

Reverse Osmosis—*continued*

Hydranautics	401 Jones Rd. Oceanside, CA 92054	Phone: (619) 901-2500 FAX: (619) 901-2578
Osmonics, Inc.	5951 Clearwater Dr. Minnetonka, MN 55343	Phone: (612) 933-2277 FAX: (612) 933-0141
Toray Industries, Inc.	Membrane Products Dept. 8-1, Mihama 1-chome, Urayasu, Chiba 279-8555, Japan	Phone: 81-047-350-6001 FAX: 81-047-350-6075
Toyobo America, Inc.	40 East 52nd St., 20th Floor New York, NY 10022	Phone: (212) 317-9245 FAX: (212) 317-9280
TriSep Corporation	93 South La Patera Lane Goleta, CA 93117	Phone: (805) 464-8003 FAX: (805) 964-1235
US Filter	10 Technology Dr. Lowell, MA 01851	Toll Free: (800) 783-PURE FAX: (508) 970-2465

Nanofiltration

The Dow Chemical Company

Fluid Systems Corporation

Hydranautics

TriSep Corporation

Appendix C

Equations

The full derivation of these equations is beyond the scope of this manual. Various manufacturers and design engineering firms use computer programs based on these and similar equations to define reverse osmosis (RO) and nanofiltration (NF) system quality and quantity. Chapter 2 of this manual provides an overview of the design considerations arising from the use of equations presented in that chapter and in this appendix.

Converting to Site-Specific Conditions

Equations C-1 and C-2 are of great practical importance in the design and construction of RO devices and plants because they may be used to convert from a manufacturer's standard conditions (SC) to site-specific operating conditions (OC).

Reverse osmosis product flow can be derived as shown in Eq C-1 (which is a minor variation of Eq 1-4):

$$Q_p = K_w A(\Delta P - \Delta\pi) \tag{C-1}$$

Where:

A	=	membrane surface area
Q_p	=	permeate flow
K_w	=	mass transfer coefficient (defined by membrane supplier)
ΔP	=	transmembrane pressure differential or gradient
$\Delta\pi$	=	osmotic pressure differential or gradient of the aqueous solution across the membrane

This equation shows that flow through the membrane is directly proportional to the applied pressure differential across the membrane minus the osmotic pressure differential.

The salt passage Q_s through the membrane is similarly derived. Salt flow is given by Eq C-2 (which is similar to Eq 1-5):

$$Q_s = K_s A \Delta C \tag{C-2}$$

Where:

Q_s = site-specific salt flow
K_s = solute flow constant defined by membrane supplier
ΔC = concentration differential across the membrane

This equation states that salt flow is proportional only to the concentration differential across the membrane. Thus, increasing the operating pressure will not affect salt flow; it will, however, based on Eq C-1, increase the water flow. Thus, it improves product quality because of the dilution resulting from the higher product flow.

Mathematically, if the operating conditions equation is divided by the standard conditions equation, two things occur. First, the C constants are canceled, making their actual value immaterial. Second, the division permits the mathematical development of a correction factor (the ratio of the OC to the SC). Based on this technique, as well as introducing a membrane aging factor (to account for fouling, compaction, and plastic creep), Eq C-3 and Eq C-4 may be developed. Equation C-3 represents flow normalization procedures; Eq C-4 shows salt passage normalization procedures.

$$Q_p = (\text{PCF})(\text{TCF})(\text{MFRC/FF})Q_i \tag{C-3}$$

$$S_p = (\text{SPCF})(\text{SP}_i) \tag{C-4}$$

Where:

Q_p = product water flow at operating conditions
TCF = temperature correction factor (generally about 3 percent per degree Celsius)
MFRC = membrane flux retention coefficient (generally about 0.65 to 0.85 over 3 to 5 years)
FF = fouling factor (generally about 0.8 to 0.9 over 3 years)
Q_i = initial product water flow at standard conditions
S_p = salt passage of RO or NF device at operating conditions
SPCF = salt passage correction factor (generally 0.5 to 1.5 of initial)
SP_i = salt passage of RO or NF device at standard conditions
PCF = pressure correction factor (not easily generalized because of osmotic pressure and bundle pressure drop effects) but can be estimated as follows:

Once the flow of an individual RO or NF device has been determined at OC, it is possible to calculate how many of these modules are required for a project:

$$\text{number of modules or elements needed} = \frac{\text{project capacity}}{Q_p} \tag{C-5}$$

$$\text{number of pressure vessels} = \frac{\text{project capacity}}{Q_p e} \tag{C-6}$$

Where:

e = number of modules or elements in a pressure vessel

Limiting Salt Calculation

This section provides equations for determining limiting salts and acid additions, and examples for their use.

$$A_nB_m = nA^{+p} + mB^{-q} \tag{C-7}$$

$$K_{sp} = [A^{+p}]^n[B^{-q}]^m \tag{C-8}$$

Where:

A_nB_m	=	salt under consideration
A^{+p}	=	cation of formed salt, *moles/L or M*
B^{-q}	=	anion of formed salt
K_{sp}	=	solubility product
p	=	charge of cation
q	=	charge of anion

Determination of Recovery from the Solubility Product

$$K_{sp} = \left[(A^{+p})^n\left(\frac{1-SP_A}{1-R}\right)\right]\left[(B^{-q})^m\left(\frac{1-SP_B}{1-R}\right)\right] \tag{C-9}$$

Where:

A^{+m}	=	feedwater concentration of cation under consideration
B^{-n}	=	feedwater concentration of anion under consideration
R	=	permeate recovery rate (expressed as decimal)
SP_A	=	passage of cation A through membrane (decimal)
SP_B	=	passage of anion B through membrane (decimal)

NOTE: SP_A and SP_B are available from membrane suppliers.

For most diffusion controlled membrane process (RO and NF) applications, SP_A and SP_B can be considered to be negligible, and the above equation simplifies to:

$$K_{sp} = \left[\frac{A^{+p}}{(1-R)}\right]^n\left[\frac{B^{-q}}{(1-R)}\right]^m \tag{C-10}$$

Example Calculations (using data from Table 2-2)

Limiting salt—$CaCO_3$

$CaCO_3 \rightarrow Ca^{+2} + CO_3^{-2}$
$K_{sp} = [Ca^{+2}]\ [CO_3^{-2}] = 10^{-8.3}$
Ca^{+2} = 8 mg/L = 0.0002 moles/L = 2×10^{-4} M
CO_3^{-2} = 3 mg/L = 0.00005 moles/L = 5×10^{-5} M

thus:

$$10^{-8.3} = \left[\frac{(2 \times 10^{-4}\text{ moles})}{X}\right]\left[\frac{(5 \times 10^{-5}\text{ moles})}{X}\right]$$

$$X = 1.41$$

or

$$R = 1 - 1.41$$

$R = -0.41$

This indicates that the solubility of $CaCO_3$ is already exceeded and $CaCO_3$ will precipitate in the feedwater.

Limiting salt—$SrSO_4$

$$SrSO_4 \rightarrow Sr^{+2} + SO_4^{-2}$$
$$K_{sp} = [Sr^{+2}]\,[SO_4^{-2}] = 10^{-6.2}$$
$$Sr^{+2} = 2 \text{ mg/L} = 0.00002 \text{ moles/L} = 2 \times 10^{-5} \text{ moles}$$
$$SO_4^{-2} = 79 \text{ mg/L} = 0.0008 \text{ moles/L} = 8 \times 10^{-4} \text{ moles}$$

thus:

$$10^{-6.2} = \left[\frac{(2 \times 10^{-5}\text{moles})}{X}\right]\left[\frac{(8 \times 10^{-4}\text{moles})}{X}\right]$$

$$X = 0.16$$

or

$$R = +1 - 0.16$$
$$R = +0.84$$

This indicates that a water recovery rate of 84% is possible before the solubility limit of $SrSO_4$ is exceeded in the feedwater.

Acid Addition to Control Calcium Carbonate Scaling

Example:

Ca^{+2} = 8 mg/L
HCO_3^- = 631 mg/L
pH = 8.0
Target product water recovery rate = 75%

Find the sulfuric acid (93 percent solution) dose necessary to prevent calcium carbonate scaling.

1. Determine required feedwater pH

$$\text{For } CaCO_3 \rightarrow Ca^{+2} + CO_3^{-2} \qquad \text{(C-11)}$$

$$K_{sp} = [Ca^{+2}]\,[CO_3^{-2}] = 10^{-8.3}$$

Where:

$$Ca^{+2} = 8 \text{ mg/L} = 0.0002 \text{ moles/L} = 2 \times 10^{-4} \text{ moles}$$
CO_3^{-2} = Unknown, must calculate from HCO_3^- concentration
$$HCO_3^- = 631 \text{ mg/L} = 0.010 \text{ moles/L} = 1 \times 10^{-2} \text{ moles}$$

Given $HCO_3^- \rightarrow H^+ + CO_3^{-2}$

$$K_2 = \frac{[H^+][CO_3^{-2}]}{[HCO_3^-]} = 10^{-10.3} \qquad \text{(C-12)}$$

or

$$CO_3^{-2} = \frac{[K_2][HCO_3^-]}{[H^+]}$$

$$CO_3^{-2} = \frac{[10^{-10.3}][0.010 \text{ moles/L}]}{[H^+]} \quad \text{(C-13)}$$

Substitute the expression for CO_3^{-2} (Eq C-13) into Eq C-7.

$$K_{sp} = \left[\frac{Ca^{+2}}{X}\right]\left[\frac{CO_3^{-2}}{X}\right]$$

$$K_{sp} = \left[\frac{2 \times 10^{-4} \text{ moles/L}}{0.25}\right]\left[\frac{(10^{-10.3})(1 \times 10^{-2} \text{ moles/L})}{(0.25H)}\right] = 10^{-8.3}$$

$$H^+ = 3.2 \times 10^{-7} \text{ moles}$$

$$\text{pH} = 6.5$$

2. Determine sulfuric acid (93% solution) dose to achieve pH of 6.5.

$$K_1 = \frac{[H^+][HCO_3^-]}{[H_2CO_3]} = 10^{-6.3} = \frac{[10^{-6.5}]\left(\left[\frac{631 \text{ mg/L}}{61,000 \text{ mg/mole}}\right] - x\right)}{[0 + X]}$$

Where:

$$X = [H^+] \text{required to react with } [HCO_3^-]$$

$$X10^{-6.3} = 3.27 \times 10^{-9} - 10^{-6.5}X$$

$$X = 0.004 \text{ moles/L } H^+ = [0.004 \text{ moles/L}]\left[\frac{2\text{EQ}}{\text{mole}}\right]\left[\frac{49,000 \text{ mg}}{\text{EQ}}\right]\left[\frac{1}{0.93}\right] = 421$$

$$X = 211 \text{ mg/L of } 93\% \text{ } H_2SO_4$$

Determination of Membrane System Configuration

This section uses an example to demonstrate how a membrane system configuration can be determined. The following assumptions were made for this example:

- spiral-wound membranes (calculation methods presented are similar for hollow fine fiber membranes except for the element surface area)
- membrane element flux = 15 gpd/ft^2
- 4-2-1 array (first stage: four pressure vessels; second stage: two; third stage: one)
- seven elements per pressure vessel (PV)
- element surface area of 350 ft^2
- recovery of 75 percent

The following calculations need to be made:

1. Water one membrane element can produce:

$$Q_p = F_w A = \left(15\frac{\text{gpd}}{\text{ft}^2}\right)\left(350\frac{\text{ft}^2}{\text{element}}\right) = 5,250\frac{\text{gpd}}{\text{element}} \quad \text{(C-14)}$$

2. Number of elements on one 4 + 2 + 1 array:

$$\left(\frac{\text{PV}}{\text{array}}\right)\left(7\frac{\text{elements}}{\text{PV}}\right) = 49\ \frac{\text{elements}}{\text{array}} \tag{C-15}$$

3. Gallons per day produced per array:

$$\left(5{,}250\frac{\text{gpd}}{\text{element}}\right)\left(49\frac{\text{elements}}{\text{array}}\right) = 257{,}250\ \frac{\text{gpd}}{\text{array}} \tag{C-16}$$

4. Arrays needed to supply 1 mgd (10^6 gpd):

$$(10^6\text{gpd})\left(\frac{1\ \text{array}}{257,\ 250\ \text{gpd}}\right) = 3.9\ \text{array (4 arrays would be used)} \tag{C-17}$$

5. Total number of elements needed:

$$(4\text{arrays})\left(49\frac{\text{elements}}{\text{array}}\right) = 196\ \text{elements} \tag{C-18}$$

6. Feedwater flow in and permeate flow out of one array:

$$\text{total feedwater flow rate} = \frac{\text{total permeate flow rate}}{\text{recovery}} = \frac{1\,\text{mgd}}{0.75} - 1.33\ \text{mgd}$$

$$\text{flow in per array} = \frac{1.33\ \text{mgd}}{4} = 0.33\ \text{mgd} = 231\ \text{gpm}$$

$$\text{flow out per array} = \frac{1\ \text{mgd}}{4} = 0.25\ \text{mgd} = 174\ \text{gpm} \tag{C-19}$$

NOTE: All arrays could be assembled into one train.

Check of Configuration on Elements

Recommended recovery per element: $3\% < R_{\text{element}} < 20\%$

1. Flow into stage 1 per PV per day

$$\left(332{,}500\ \frac{\text{gal}}{\text{array day}}\right)\left(7\frac{\text{array}}{4\ \text{PV}}\right) = 83{,}125\frac{\text{gal}}{4\ \text{PV day}}\ \text{for stage 1} \tag{C-20}$$

2. Permeate flow per element per day assuming a flux of 15 g/ft^2/d

$$\left(350\frac{\text{ft}^2}{\text{element}}\right)\left(15\frac{\text{gal}}{\text{ft}^2/\text{day}}\right) = 5,\ 250\frac{\text{gal}}{\text{element-day}} \tag{C-21}$$

3. Recovery in first and last element of a PV in stage 1

$$R_{E11} = \frac{Q_p}{Q_f} = \frac{5{,}250\text{ gal}}{83{,}125\text{ gal}} = 0.06\text{ or }6\%$$

$$R_{E17} = \frac{Q_p}{Q_f} = \frac{5{,}250\text{ gal}}{83{,}125\text{ gal} - [(7-1\text{ elements}) \times (5{,}250)]\text{ gal}} = 0.09\text{ or }9\% \tag{C-22}$$

4. Recovery in first and last element of a PV in stage 2

Flow into a PV in stage 2 = 2[83,125 gal – 7(5,250)] = 92,750 gal

$$R_{E11} = \frac{Q_p}{Q_f} = \frac{5{,}250\text{ gal}}{(92{,}750\text{ gal})} = 0.06\text{ or }6\%$$

$$R_{E17} = \frac{Q_p}{Q_f} = \frac{5{,}250\text{ gal}}{92{,}750\text{ gal} - (6 \times 5{,}250)\text{ gal}} = 0.09\text{ or }9\% \tag{C-23}$$

5. Recovery in first and last element of a PV in stage 3

Flow into a PV in stage 3 = 2[92,750 gal – 7(5,250)] = 112,000 gal

$$R_{E11} = \frac{Q_p}{Q_f} = \frac{5{,}250\text{ gal}}{112{,}000\text{ gal}} = 0.05\text{ or }5\%$$

$$R_{E17} = \frac{Q_p}{Q_f} = \frac{5{,}250\text{ gal}}{112{,}000\text{ gal} - (6 \times 5{,}250)\text{ gal}} = 0.07\text{ or }7\% \tag{C-24}$$

Posttreatment Equations

Alkalinity (all species in moles per litre):

$$\text{Alkalinity} \leftrightharpoons 2(CO_3^{2-}) + (HCO_3^-) + (OH^-) - (H^+) \tag{C-25}$$

Calcium carbonate precipitation:

$$CaCO_3(s) \leftrightharpoons Ca^{2+} + CO_3^{2-} \tag{C-26}$$

Solubilization of carbon dioxide gas:

$$CO_{2(g)} \leftrightharpoons CO_{2(aq)} \tag{C-27}$$

$$pK_H = 1.5$$

$$CO_{2(aq)} + H_2O \leftrightharpoons H_2CO_3$$

$$pK_m = 2.8$$

Where:

pK_H = $-\log K_1$
pK_m = equilibrium constant

Ionization of aqueous carbonic acid:

$$H_2CO_3^- \leftrightharpoons H^+ + HCO_3^{2-} \tag{C-28}$$

$$pK_1 = 6.3$$

Ionization of bicarbonate:

$$H_2CO_3^- \leftrightharpoons H^+ + HCO_3^{2-} \tag{C-29}$$

$$pK_2 = 10.3$$

Caustic stabilization:

$$HCO_3^- + H^+ + HCO_3 \tag{C-30}$$

$$H_2CO_3 \rightarrow CO_{2(g)} + H_2O \tag{C-31}$$

Alkalinity recovery:

$$CO_{2(2)} + OH^- \rightarrow HCO_2^- \tag{C-32}$$

Elemental sulfur turbidity production (inefficient hydrogen sulfide stripping):

$$2H_2S + O_2 = 2H_2O + 2S_{(s)} \tag{C-33}$$

Hydrogen sulfide reactions with water:

$$H_2S + H_2O = H_3O^+ + HS^-$$

$$pK_1 = 7.0 \tag{C-34}$$

Where:

pK_1 = ionization constant 1

and

$$HS^- + H_2O = H_3O^+ + S^{2-}$$

$$pK_2 = 14.0 \tag{C-35}$$

Where:

pK_2 = ionization constant 2

Hypochlorous acid formation:

$$Cl_2 + H_2O = HOCl + H^+ + Cl^- \tag{C-36}$$

$$K_{a,1} = \frac{[H^+][Cl^-][HOCl]}{[Cl_2]} = 4 \times 20^{-4} \text{ (at 77° F)}$$

Where:

$K_{a,1}$ = equilibrium constant for acid ionization

Hypochlorous acid dissociation:

$$HOCl = H^+ + OCl^- \tag{C-37}$$

$$K_{a,2} = \frac{[OCl][H^+]}{[HOCl]} = 2.7 \times 10^{-a} \text{ (at 68° F)}$$

Where:

$K_{a,2}$ = equilibrium constant for acid ionization

Chlorine reactions with hydrogen sulfide:

$$H_2S + HOCl = HCl + H_2O + S_{(s)} \text{ for pH} > 3.8 \tag{C-38}$$

$$H_2S + 4HOCl = 4HCl + H_2SO_4 + S_{(s)} \text{ for pH} > 3.8 \tag{C-39}$$

pH of saturation:

$$pH_s = -\log \frac{K_2}{K_{sp}}[Ca^{2+}][Alk] \tag{C-40}$$

Where:

K_2 = H_2CO_3 second ionization constant
K_{sp} = $CaCO_3$ solubility product constant
Ca^{2+} = calcium ion concentration, in moles/L
Alk = alkalinity, in moles/L

Lime stabilization:

$$CO_2 + Ca(OH)_2 = CaCO_3 + H_2O \tag{C-41}$$

Calcium carbonate stabilization:

$$CO_2 + CaCO_3 + H_2O = Ca^{2+} + 2HCO_3^- \tag{C-42}$$

Caustic stabilization:

$$CO_2 + 2NaOH = 2Na^+ + CO_3^{2-} + H_2O \tag{C-43}$$

Soda ash stabilization:

$$Na_2CO_3 = 2Na^+ + CO_3^{2-} \tag{C-44}$$

This page intentionally blank.

Appendix D

Silt Density Index Procedure

The equipment required to perform the test to determine the silt density index (SDI) includes the SDI apparatus (shown schematically in Figure D-1), a 500-mL graduated cylinder, and a stopwatch. The procedure is as follows:

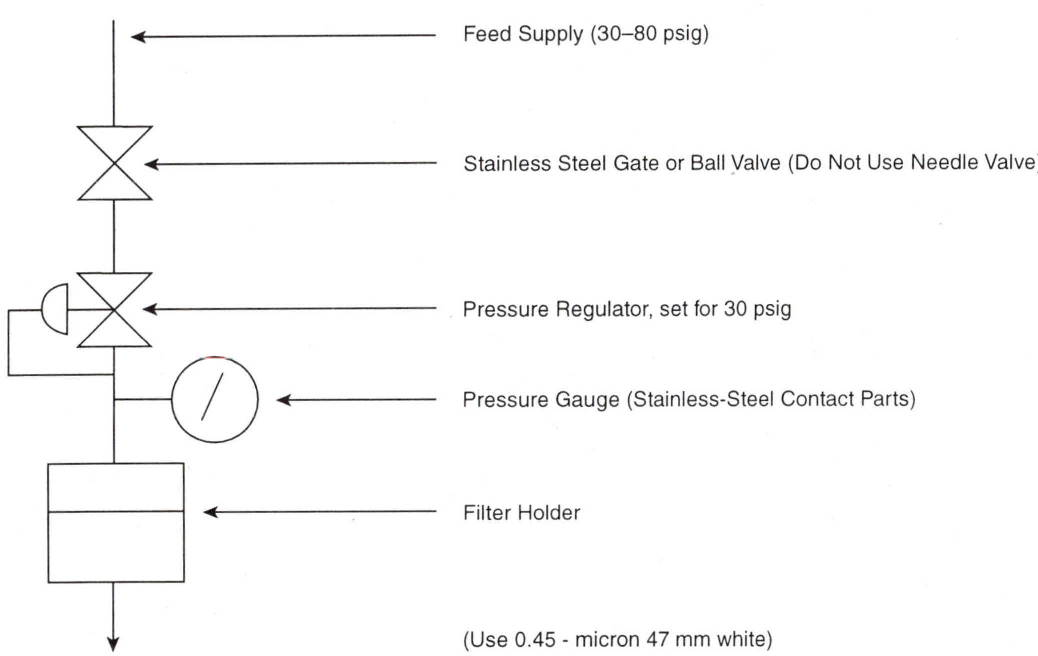

Figure D-1 Silt density index apparatus

1. Using the apparatus shown schematically in Figure D-1, set the pressure regulator to 30 psig.
2. Flush the sample line to be sure it is free from accumulated debris that would interfere with filtration before installing the 0.45-micron membrane filter disk. Allow the water to run long enough to establish a constant temperature.
3. Open the filter holder and insert the membrane filter. Avoid handling the membrane filter with fingers. It should be handled only with dull tweezers around the edges to avoid punctures and contamination.
4. Reposition the O-ring and close the top half of the filter holder, but do not tighten the wing nuts.
5. Open the ball valve just enough to allow trapped air to bleed from the apparatus and water line. Close the ball valve and tighten the filter holder.
6. Open the ball valve, and with a stopwatch immediately begin to measure the time required to collect 500 mL of filtrate. Record this time (t_i). Allow water to continue to flow through the filter under constant pressure (30 psig).
7. After 15 minutes of total elapsed time, measure and record the time (t_f) required to collect another 500 mL of filtrate.
8. The SDI is calculated from the following equation:

$$SDI = \frac{100\left[1-\left(\frac{t_i}{t_f}\right)\right]}{t_t} \qquad \text{(D-1)}$$

Where:

t_i = time to collect the initial 500 mL of filtrate
t_f = time to collect 500 mL of filtrate after the filter has been on-line for a period of time, usually 15 minutes
t_t = total test time, usually 15 minutes

Appendix E

Langelier Saturation Index and Stiff and Davis Scaling Index

The Langelier Saturation index (LSI), as modified by Larson and Buswell (1944), is defined as follows:

$$LSI = pH - pCa - pAlk - C \qquad \text{(E-1)}$$

Where:

pCa = negative logarithm of the calcium molarity
pAlk = negative logarithm of the alkalinity in equivalents per litre
C = constant to account for the change in calcium carbonate solubility with temperature and ionic strength (Figure E-1)

Consider water that has the following attributes:

- pH of concentrate stream (by measurement), 8.5
- calcium ion concentration, 150 mg/L
- total alkalinity as calcium carbonate, 300 mg/L
- total dissolved solids, 2,000 mg/L

The individual terms in Eq E-1 are

$$pCa = -\log\left(\frac{50 \text{ mg/L as } CaCO_3}{40{,}000 \text{ mg/mole}}\right) = 2.42$$

$$pAlk = \log\left(\frac{300 \text{ mg/L as } CaCO_3}{40{,}000 \text{ mg/equiv.}}\right) = 2.22$$

Where:

C	=	2.34, from Figure E-1
LSI	=	8.5 – 2.42 – 2.22 – 2.34 = +1.52

The Stiff and Davis Scaling Index (SDSI) (Stiff and Davis 1952) is defined as follows:

$$SDSI = pH - pCa - pAlk - K \qquad \text{(E-2)}$$

Where:

pCa	=	negative logarithm of the calcium molarity, as in LSI
pAlk	=	negative logarithm of the alkalinity in equivalents per litre, as in LSI
K	=	different empirical constant from LSI to account for temperature and ionic strength (Figure E-2)

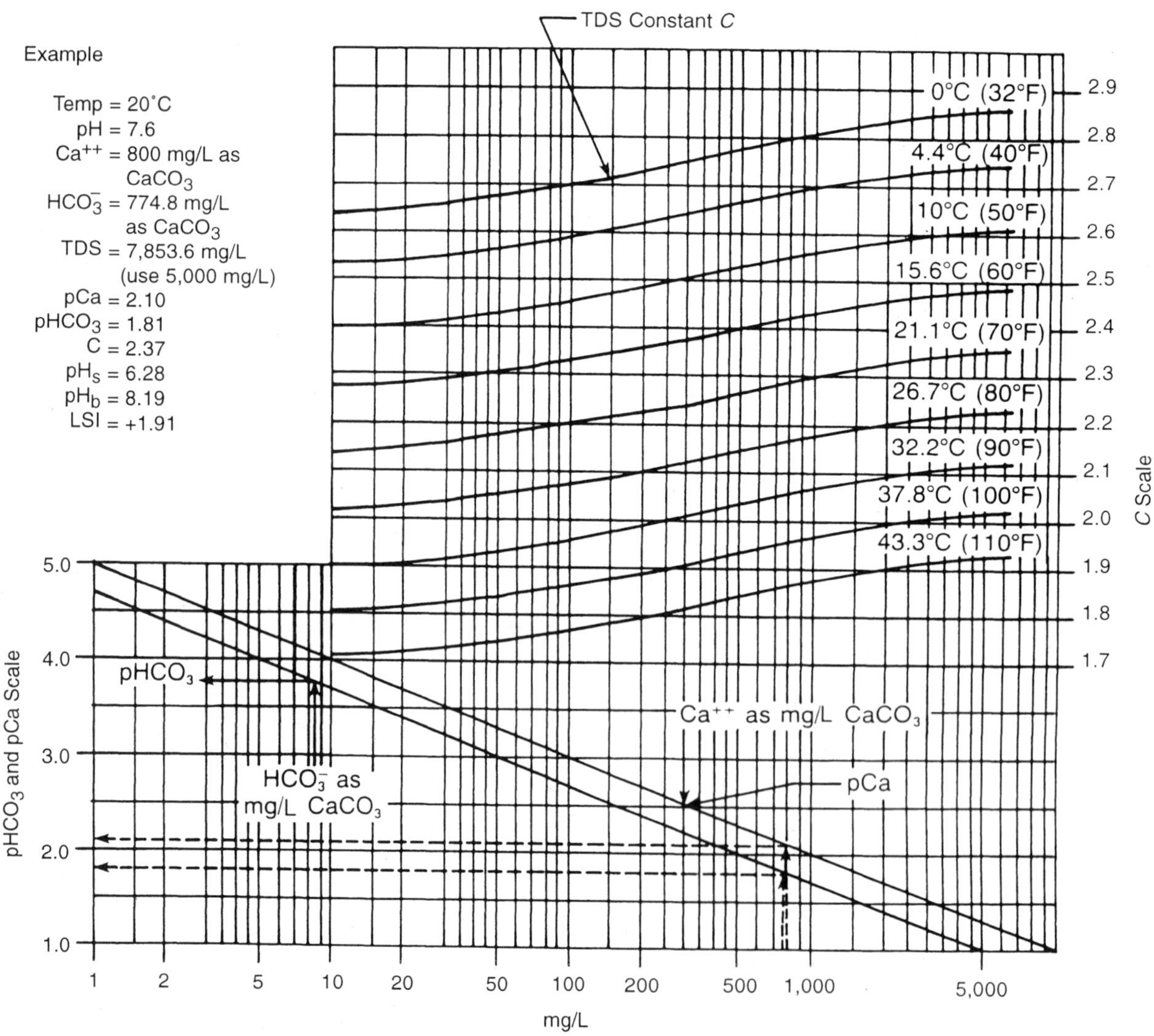

Figure E-1 Nomograph for calculating the Langelier Saturation index

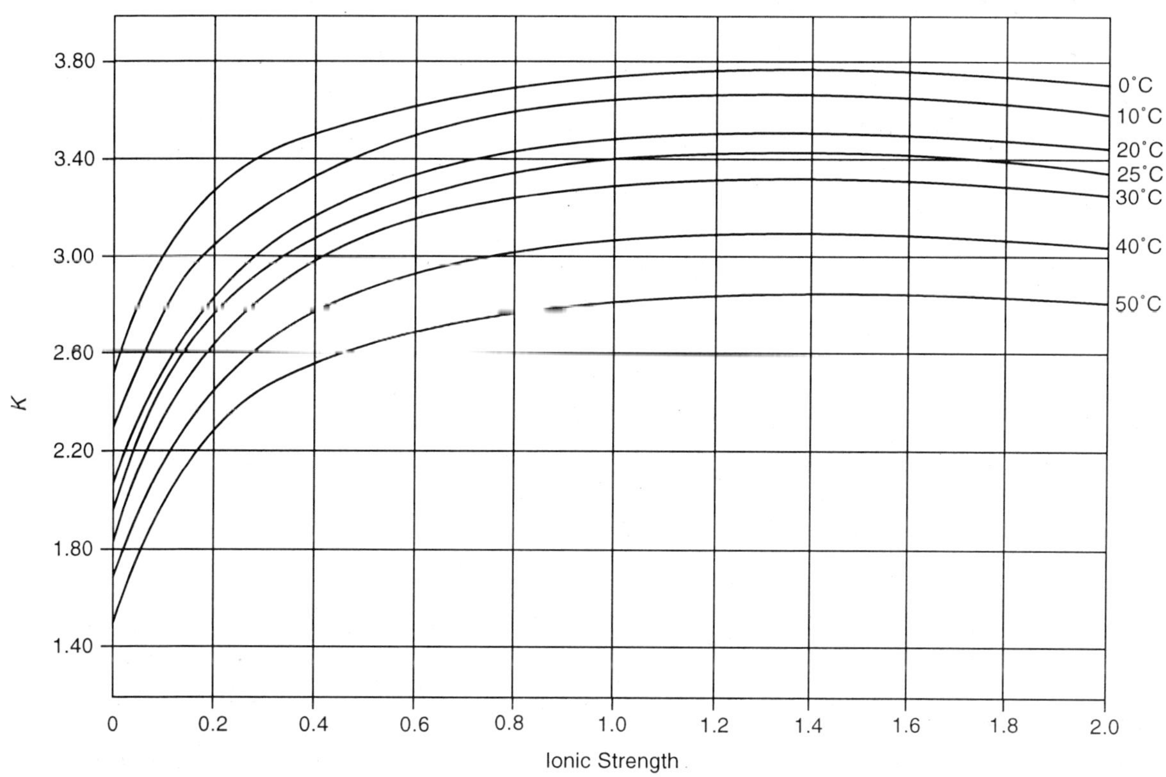

Figure E-2 Stiff and Davis K versus ionic strength and temperature

REFERENCES

Larson, T.E., and A.M. Buswell. 1944. Calcium Carbonate Saturation Index and Alkalinity Interpretations. *Jour. AWWA,* 34:1667.

Stiff, H.A., and L.E. Davis. 1952. *A Method for Predicting the Tendency of Oil Field Waters to Deposit Calcium Carbonate.* Petroleum Transactions, AIME.

This page intentionally blank.

Additional Sources of Information

Chapter 1

Alabdula'aly, A.I. 1995. Membrane Process Application in the Central Region of Saudi Arabia. In *Proc. AWWA, Membrane Technology Conference*. Denver, Colo.: American Water Works Association.

American Society for Testing and Materials (ASTM). 1982. *Test Method for Silt Density Index.* Standard D4189-82. Philadelphia, Pa.: ASTM.

———. 1989. *Guide for Recordkeeping for Reverse Osmosis Systems.* Standard D4472-89. Philadelphia, Pa.: ASTM.

———. 1989. *Practice for Calculation and Adjustment of Silica (SiO_2) Scaling for Reverse Osmosis.* Standard D4993-89. Philadelphia, PA.: ASTM.

———. 1989. *Practice for Detecting Leaks in Reverse Osmosis Devices.* Standard D3923-80. Philadelphia, Pa.: ASTM.

———. 1989. *Practice for Standardizing Reverse Osmosis Performance Data.* Standard D4516-85. Philadelphia, Pa.: ASTM.

———. 1989. *Test Methods for Operating Characteristics for Reverse Osmosis Devices.* Standard D4194-89. Philadelphia, Pa.: ASTM.

———. 1991. *Practice for Calculation and Adjustment of the Stiff and Davis Stability Index for Reverse Osmosis.* Standard D4582-91. Philadelphia, Pa.: ASTM.

———. 1992. *Practice for Calculation and Adjustment of Sulfate Scaling Salts ($CaSO_4$, $SrSO_4$, $BaSO_4$) for Reverse Osmosis.* Standard D4692-87. Philadelphia, Pa.: ASTM.

———. 1993. *Guide for Water Analysis for Reverse Osmosis Applications.* Standard D4195-88. Philadelphia, Pa.: ASTM.

———. 1993. *Practice for Calculation and Adjustment of the Langelier Saturation Index for Reverse Osmosis.* Standard D3739-88. Philadelphia, Pa.: ASTM.

———. 1994. *Test Method for Silt Density Index (SDI) of Water.* Standard D4189-94. Philadelphia, Pa.: ASTM.

———. 1995. *Test Method for Operating Characteristics of Home Reverse Osmosis Devices.* Standard D5615. Philadelphia, Pa.: ASTM.

Applegate, L.E. 1984. Membrane Separation Processes. *Chem Eng.* June 11;64–80.

Belfort, G. 1984. *Synthetic Membrane Processes.* New York: Academic Press.

Bergman, R.A., and J.C. Lozier. 1990. Expanding Applications for Membrane Processes in Water Treatment. In *Proc. of the National Water Supply Improvement Association Biennial Conference.* August 19–23. Lake Buena Vista, Fla.: National Water Supply Improvement Association.

Bergman, R.A., and J.C. Lozier. 1993. Membrane Process Selection and the Use of Bench and Pilot Tests. In *Proc. of the AWWA Membrane Technology Conference.* Denver, Colo.: American Water Works Association.

Chapman-Wilbert, M. 1993. *The Desalting and Water Treatment Membrane Manual: A Guide to Membranes for Municipal Water Treatment.* Water Treatment Technology Program Report No. 1. Denver, Colo.: US Department of Interior, Bureau of Reclamation, Denver Office. Research and Laboratory Services Division. R-93-15.

Computer Projection Programs and Engineering Manuals by:

- Desalination Systems, Inc. Vista, Calif.
- Dow Chemical Company, Filmtec Corporation. Midland, Mich.
- DuPont Company, Permasep® Products. Newark, Del.
- Fluid Systems Corporation. San Diego, Calif.
- Hydranautics. Oceanside, Calif.
- Osmonics, Inc. Minnetonka, Minn.
- TriSep Corporation. Goleta, Calif.

Conlon, W.J., and S.A. McClellan. 1989. Membrane Softening, A Water Treatment Process That Has Come of Age. *Jour. AWWA*, 81(11):47–51.

Duranceau, S.J. 1994. Mathematics for Membrane Processes. *Florida Water Resources Journal*, 46(9):26–28.

Duranceau, S.J., J.S. Taylor, and L.A. Mulford. 1992. SOC Removal in a Membrane Softening Process. *Jour. AWWA*, 84(1):68–78.

Freeman, S.D.N., and J. Crook. 1995. An Update on Membrane Water Reuse Projects. In *Proc. AWWA Membrane Technology Conference*. Denver, Colo.: American Water Works Association.Jacangelo, J.G., and J.M. Laine. 1994. *Evaluation of Ultrafiltration Membrane Pretreatment and Nanofiltration of Surface Waters*. Denver, Colo.: American Water Works Association Research Foundation and American Water Works Association.

Kedem, O., and A. Katchalsky. 1958. Thermodynamic Analysis of the Permeability of Biological Membranes to Non-Electrolytes. *Biochimica et Biophisica Acta*, 27:123.

Kruithof, J.C., J.A.M.H. Hofman, and R. Hopman. 1995. Rejection of Pesticides and Other Micropollutants by Reverse Osmosis. In *Proc. AWWA Membrane Technology Conference*. Denver, Colo.: American Water Works Association.

Laisure, D., L. Sung, and J.S. Taylor. 1994. *Fundamental Membrane Phenomena. Desalination and Water Reuse*, 3(3):10–13.

Pontius, F.W., ed. 1990. *Water Quality and Treatment,* 4th ed. Denver, Colo.: American Water Works Association.

Rautenbach, R., and R. Albrecht. 1989. *Membrane Processes*. New York: John Wiley & Sons.

Riley, R.L. 1990. Reverse Osmosis. *Membrane Separation Systems,* 5(1):5–52. US Department of Energy. Published by Department of Commerce, National Technical Information Service.

Sourirajan, S. 1977. *RO and Synthetic Membranes.* Ottawa, Ont.: NRC.

Sourirajan, S., and Matsurra, T. 1985. *Reverse Osmosis/Ultrafiltration Process Principles*. Division of Chemistry. Ottawa, Ont.: NRC.

Sung, L. August 1993. *Film-Theory and Ion Coupling Models for Diffusion Controlled Membrane Processes.* PhD diss. Orlando, Fla.: University of Central Florida.

Taylor, J.S., L.A. Mulford, W.M. Barrett, S.J. Duranceau, and D.K. Smith. 1989. *Cost and Performance of Membrane Processes for Organic Control on Small Systems.* EPA/600/S-2/89/022. Cincinnati, Ohio: USEPA Risk Reduction Engineering Research Laboratory.

Water Treatment Handbook, 6th ed. 1991. Cedex, France: Degremont.

Chapter 2

Allgeier, S.C., and R.S. Summers. 1996. Development of a Rapid Bench-Scale Membrane Test for the Evaluation of Nanofiltration Performance. In *Proc. AWWA Membrane Technology Conference*. Denver, Colo.: American Water Works Association.———. 1996. Effect of Mass Transfer Resistance and System Recovery on Membrane Permeation. In *Proc. AWWA Membrane Technology Conference*. Denver, Colo.: American Water Works Association.

American Water Works Association. 1997. *Water Wells.* Standard A100-97. Denver, Colo.: AWWA.

Applegate, L.E. 1984. Membrane Separation Processes. *Chemical Engineering,* 6:11.

——— 1986. Post-Treatment of Reverse Osmosis Product Waters. *Jour. AWWA*, 78(5):59.

Argo, D.G. Demineralization. In *Operation of Water Treatment Plants*, Vol. II. Sacramento, Calif.: California State University.

Beardsley, S.S. 1991. Potable Water Post Treatment. In *Proc. AWWA Membrane Technologies for the Water Industry Specialty Conference.* Denver, Colo.: American Water Works Association.

Belfort, G. 1984. *Synthetic Membrane Processes*. Orlando, Fla.: Academic Press.

Blau, T.J., J.S. Taylor, K.E. Morris, and L.A. Mulford. 1992. DBP Control by Nanofiltration: Cost and Performance. *Jour. AWWA*, 84(12):104–116.

Blind, R., and R. Hollander. 1972. *Hydrogen Sulfide Removal from Reverse Osmosis Product Water.* Patent 5,096,589. International Process Technology, Inc., West Palm Beach, Fla.

Conlon, W.J. 1985. Pilot Plant Test Data for Prototype Ultra-Low Pressure Reverse Osmosis Elements. In *Proc. Second World Congress on Desalination and Water Reuse.* Vol. 3. Boca Raton, Fla.: Lineal Publishing Company.

Conlon, W.J., and S.A. McClellan. 1989. Membrane Softening, A Water Treatment Process That Has Come of Age. *Jour. AWWA*, 81(11):47–51.

Cornwell, D.A. 1990. Air Stripping and Aeration. In *Water Quality and Treatment.* F.W. Pontius, ed. New York: McGraw-Hill Book Company.

Duranceau, S.J., J.S. Taylor, and L.A. Mulford. 1992. SOC Removal in a Membrane Softening Process. *Jour. AWWA*,

84(1):68–78.Driksson, P. 1988. Nanofiltration Extends the Range of Membrane Filtration. *Environ. Progress*, 7(1):158–162.

Envirogenics Handbook for Engineering of Reverse Osmosis Systems. circa 1974. El Monte, Calif.: Envirogenics Systems Company.

Gekkas, V. 1988. Terminology for Pressure-Driven Membrane Operations. *Desalination*, 68:77–92. Amsterdam, Netherlands: Elsevier Science.

Hackett, G., and J.H. Lehr. 1985. *Iron Bacteria: Occurrence, Problems, and Control Methods in Water Wells.* Worthington, Ohio: National Water Well Association.

Harris, F.L., G.B. Humphreys, and K.S. Spiegler. 1976. Reverse Osmosis (Hyperfiltration) in Water Desalination. In *Membrane Separation Processes.* P. Meares, ed. Amsterdam, Netherlands: Elsevier Science.

Hem, J.D. 1985. *Study and Interpretation of the Chemical Characteristics of Natural Water.* Reston, Va.: US Geological Survey Water-Supply Paper 2254.

Hollosep Technical Bulletin. Reverse Osmosis Department. Osaka, Japan: Toyoba Company, Ltd.

Lohman, S.W. 1979. *Ground-Water Hydraulics.* Reston, Va.: US Geological Survey Professional Paper 708.

Lyn, T.L., and J.S. Taylor. 1992. Assessing Sulfur Turbidity Formation Following Chlorination of Hydrogen Sulfide in Groundwater. *Jour. AWWA*, 82(9):103.

Mallevaille, J., P.E. Odendall, and M.R. Wiesner, eds. 1996. *Water Treatment Membrane Processes.* New York: McGraw-Hill Book Company.

Mickley, M., R. Hamilton, L. Gallegos, and J. Truesdall. 1993. *Membrane Concentrate Disposal.* Denver, Colo.: American Water Works Association Research Foundation and American Water Works Association.

Permasep Products Engineering Manual. 1992. Wilmington, Del.: DuPont Company.

Pohland, H.W. 1987. Reverse Osmosis Membranes. In *Proc. of the NWSIA and SFWMD Seminar.*

Pontius, F.W., J. Catlin, E. Kawczynski, and S.J. Koorse. 1996. Federal and State Regulations Governing Membrane Concentrate Disposal. In *Proc. AWWA Membrane Technology Conference.* Denver, Colo.: American Water Works Association.

Reverse Osmosis Engineering Newsletters. 1981. El Monte, Calif.: Envirogenics Systems Company.

Sawyer, C.N., and P.L. McCarty. 1978. *Chemistry for Environmental Engineering.* New York: McGraw-Hill Book Company.

Schweitzer, P.A., ed. 1988. *Handbook of Separation Techniques for Chemical Engineers.* 2nd ed. Sections 2.1 and 4.6. A James Peter Book. New York: McGraw-Hill Book Company.

Souriraian, S., and T. Matsura. 1985. *Reverse Osmosis/Ultrafiltration Process Principles.* Division of Chemistry. Ottawa, Canada: National Research Council of Canada.

Sudak, R.G. 1990. Reverse Osmosis. In *Handbook of Industrial Membrane Technology.* M.C. Porter, ed. Park Ridge, N.J.: Noyes Publications.

Suratt, W.B. 1993. Pretreatment for Membrane Processes. In *Proc. AWWA Membrane Technology Conference.* Denver, Colo.: American Water Works Association.

A Study of Reverse Osmosis. circa 1976. San Diego, Calif.: Fluid Systems Division, UOF, Inc.

Tan, L., and G.L. Amy. 1991. Comparing Ozone and Membrane Separation for Color Removal and Disinfection By-Product Control. *Jour. AWWA*, 83(3):74–79.

Taylor, J.S., Y.A. Yousef, L.A. Mulford, and T.L. Lyn. 1992. Corrosion Control of Finished Water from a Membrane Process. In *Proc. AWWA Annual Conference and Exhibition.* Denver, Colo.: American Water Works Association.

Taylor, J.S., et al. 1986. *Cost and Performance Evaluation of In-Plant Trihalomethane Control Techniques.* EPA/600/52–85/138. Cincinnati, Ohio: US Environmental Protection Agency.

Taylor, J.S., D.M. Thompson, and J.K. Carswell. 1987. Applying Membrane Processes to Groundwater, Sources for THM Precursor Control. *Jour. AWWA*, 79(8):72–82.

Taylor, J.S., L.A. Mulford, S.J. Duranceau, and W.M. Barrell. 1989. Cost and Performance of Membrane Pilot Plant. *Jour. AWWA*, 81(11):52–60.

Taylor, J.S., et al. 1989. *Cost and Performance of Membrane Processes for Organic Control in Small Systems.* EPA/600/S-2/2/89/002. Cincinnati, Ohio: US Environmental Protection Agency.

Todd, D.K. 1980. *Groundwater Hydrology*, 2nd ed. New York: John Wiley and Sons.

Treybal, R.D. 1968. *Mass Transfer Operations.* New York: McGraw-Hill Publishers.Vigneswaran, S., and R.B. Airn, eds. 1989. *Water, Wastewater, and Sludge Filtration.* Boca Raton, Fla.: CRC Press.

Vigneswaran, S., and R.B. Airn, eds. 1989. *Water, Wastewater, and Sludge Filtration.* Boca Raton, Fla.: CRC Press.

White, G. 1972. *Handbook of Chlorination.* New York: Van Nostrand Reinhold Company.

Chapter 3

Burns and Roe Industrial Services Corporation. 1979. *Reverse Osmosis Technical Manual.* US Department of Interior, Office of Water Research and Technology. US Department of National Technical Information Service. Springfield, Va.: US Department of Commerce.

Code Line Pressure Vessels. 1990. Product Bulletin 507011A, User's Guide. San Marcos, Calif.: Advanced Structures, Inc.

Element Loading Instructions. San Diego, Calif.: Fluid Systems Corporation.

Hydraulic Institute Standards for Centrifugal, Rotary, and Reciprocating Pumps, 14th ed. 1993. Cleveland, Ohio: Hydraulic Institute.

James M. Montgomery Consulting Engineers, Inc., 1985. *Water Treatment Principles and Design.* New York: John Wiley and Sons, Inc.

Permasep Engineering Manual. 1992. Section IV, RO Pretreatment, Operations, and Maintenance. Wilmington, Del.: DuPont Company.

Oreskovich, R. 1992. Utility's Viewpoint in Membrane Facility Design. In *Proc. AWWA Annual Conference and Exposition.* Denver, Colo.: American Water Works Association.

PWT Americas, Inc. *Internal Guidelines for Equipment Installation.*

Whitney, D.C. 1995. Funding: Key Ingredient to a Small Community Membrane System. In *Proc. AWWA Membrane Technology Conference.* Denver, Colo.: American Water Works Association.

Wilbert, M.C., F. Leitz, E. Abart, B. Boegli, and K. Linton. 1998. *The Desalting and Water Treatment Manual: A Guide to Membranes for Municipal Water Treatment,* 2nd ed. Water Treatment Technology Program Report No. 29. Denver, Colo.: Bureau of Reclamation, US Department of the Interior.

Chapter 4

Constock, D.L. 1991. Reduction of Colloidal Fouling in Membrane Systems by Organic Antifoulant Formulations. *Industrial Water Treatment,* 23(4):39–42.

Durham, L.A. 1989. Biological Growth Control in RO Systems. *Ultrapure Water,* 6(6):30–37.

———. 1991. Qualifying Colloidal Fouling in RO Systems. *Ultrapure Water,* October: 65–67.

Glossary

acidity The quantitative capacity of aqueous media to react with hydroxyl ions.

alkaline scale Scale that will dissolve under acidic conditions; usually composed of calcium carbonate and magnesium hydroxide.

alkalinity A measure of the ability of a water to neutralize acids; the sum of titratable bases. Bicarbonate, carbonate, and hydroxides in natural or treated water are major contributors to alkalinity

alum Aluminum sulfate, $Al_2(SO_4)_3 \bullet 18H_2O$.

ambient temperature The temperature of the surroundings, usually taken as 70°F.

amorphous Noncrystalline; lacking any regular cohesive structure.

anion The ion in an electrolytic solution that migrates to the anode. It carries a negative charge.

anode The positive electrode of an electrodialysis cell.

angstrom A unit of length equivalent to 10^{-10} metres, 10^{-4} microns, 10^{-8} centimetres, and 4×10^{-9} in., indicated by the symbol Å, A, or A.U.

antiscalant A chemical that inhibits scale formation.

antitelescoping device A plastic cover, resembling a wheel with spokes, attached to the ends of a spiral-wound cartridge to prevent movement of the cartridge leaves in the feed flow direction due to high feed flows.

AOC Assimilable organic carbon.

aquifer A geological formation, group of formations, or part of a formation capable of yielding a significant amount of water to a well or spring.

aramid An aromatic polyamide.

array A series of installed pressure vessels with common feedwater, product, and concentrate lines.

ATP Adenosine triphosphate.

autopsy The dissection of a membrane element to investigate causes for unsatisfactory performance.

AWWA American Water Works Association.

AWWARF American Water Works Association Research Foundation.

backwash The process of reversing the flow of water either across or through a medium or a membrane.

bacteria Any of a class of microscopic single-celled organisms that reproduce by fission or by spores. Bacteria are characterized by round, rod-like, spiral, or filamentous bodies, often in colonies or moving by means of flagella. They are widely dispersed in soil, water, organic matter, and the bodies of plants and animals. They are often symbiotic in humans, but sometimes pathogenic.

bactericide An agent capable of destroying bacteria.

bacteriostat A substance that inhibits bacterial growth and metabolism.

bank A grouping of modules and a high-pressure pump.

BAT Best available technology for a particular contaminant as defined by the US Environmental Protection Agency.

biological deposits Deposits of organisms or the products of their life processes.

biomass Any material that is or was part of a living organism.

blinding In depth and surface filtration, a buildup of particulates on or within the filter, preventing fluid flow through the filter at normal pressures.

boundary layer A very thin layer adhering to a membrane facing the feedwater or concentrate water stream.

brackish water Water having a total dissolved solids concentration ranging from 1,000 to 30,000 mg/L.

brine A concentrate stream containing total dissolved solids at a concentration greater than 36,000 mg/L.

brine seal A rubber lip seal on the outside of a spiral-wound cartridge that prevents feed bypass between the cartridge and the inside pressure vessel wall.

bundle A collection of parallel filaments or fibers.

cathode The negative electrode of an electrodialysis cell.

cation A positively charged ion in solution that migrates to the cathode.

cellulose The carbohydrate that is the principal constituent of wood.

cellulose acetate A polymer used to make semipermeable membranes.

channeling A condition of unequal flow distribution in a desalination bundle or filter bed.

chelating agent A sequestering or completing agent that, in aqueous solution, renders a metallic ion inactive through the formation of an inner ring structure with the ion.

chemical rejuvenation Any of several in-place chemical cleaning methods to remove fouling and scaling or to recondition membranes.

CIP Cleaning in place.

compaction Compression of reverse osmosis membranes due to long-term exposure to pressure resulting in a decreased water flux.

composite membrane A membrane obtained by precipitating a thin desalinating layer on a porous carrier membrane.

concentrate The membrane output stream that contains water rejected by the membrane. It is where feedwater constituents are concentrated. It is also known as reject, retentate, or the residual stream.

concentration polarization The phenomenon of increased salt concentration relative to the bulk solution that occurs in a thin boundary layer at a membrane surface on its high-pressure side.

contaminant Any foreign substance present that will adversely affect performance.

control block A group of permeators having a common piping and control system.

cpu Chloroplatinate units (color indicator).

DBP Disinfection by-product.

D–DBP Rule The Disinfectant–Disinfection By-Product Rule.

decarbonation A process to remove carbon dioxide in the form of CO_2 gas from feedwater.

degasification The process of removing dissolved gases from water.

demineralization The process of removing minerals from water, usually through deionization, reverse osmosis, or distillation.

DOC Dissolved organic carbon.

double-pass RO system A reverse osmosis system in which the permeate is further processed by a subsequent reverse osmosis system.

drawback The reverse flow of permeate from the permeate side across the membrane to the feedwater or concentrate side as a result of osmosis.

electrodialysis Dialysis conducted with the aid of an electromotive force applied to electrodes adjacent to both sides of the membrane.

ESWTR Enhanced Surface Water Treatment Rule.

feed channel spacer A polypropylene netting between membrane leaves that increases the turbulence of the feed-brine stream.

feed distributor The plastic mesh cylinder found at the core of the fiber bundle that distributes the feed evenly.

feedwater Influent or source water into the membrane process.

fiber bundle The heart of a permeator, consisting of a hollow fiber polymer membrane, epoxy tube sheet, nub, and feed distributor.

filtrate The portion of the feedwater that has passed through a filter.

flat sheet membrane A reverse osmosis membrane coated onto a fabric substrate.

flux of water (F_w) The rate of water flow across the membrane surface area, (Q_p/A), typically expressed in gallons per day per square foot (commonly abbreviated as gpd/ft^2).

fouling A reduction in water mass transfer by materials in the water, typically caused by silts and colloids.

FRP Fiberglass-reinforced plastic.

GAC Granular activated carbon.

gpd Gallons per day.

groundwater Water confined in permeable sand layers between rock or clay; that part of the subsurface water that is in the saturated zone.

HAA Haloacetic acid.

hardness The concentration in water of polyvalent cations, generally calcium and magnesium.

hard scale Deposits of calcium sulfate or other materials that cannot be dissolved by acid.

HSD Homogenous solution diffusion.

infiltration The movement of water into and through a soil. IOC. Inorganic chemical.

ion An electrified portion of matter of atomic or molecular dimensions.

ionic strength A measure of the overall electrolytic potential of a solution.

Langelier Saturation index A calculated value based on total dissolved solids, calcium concentration, total alkalinity, pH, and solution temperature. This index shows the tendency of a water solution to precipitate or dissolve calcium carbonate.

leaf A combination of a flat sheet membrane, a product channel spacer, and another flat sheet membrane, layered and glued together on three sides.

lime Calcium oxide, CaO. limestone. Either calcite limestone ($CaCO_3$) or dolomitic limestone ($CaCO_3$ and $MgCO_3$).

LSI *See* Langelier Saturation index. mass transfer. The passage of a given mass of material through a membrane to the permeate side.

mass transfer coefficient (MTC) A coefficient quantifying material passage though a membrane. The MTC of water is called K_w, and the solute MTC is K_s.

MCL. Maximum contaminant level.

membrane A highly engineered polymer film containing controlled distributions of pores. Membranes serve as a barrier permitting the passage of materials only up to a certain size, shape, or character. Membranes are used as a separation mechanism in water treatment, laboratory, and industrial applications. membrane compaction. *See* compaction.

membrane configuration The arrangement of individual elements (cartridges) in a membrane treatment process. membrane element. A single membrane unit or cartridge.

membrane system Several membrane trains in parallel. microfiltration (MF). Filtration designed to remove particles and bacteria in the approximate range of 0.05 to 10 micrometres.

milliequivalents per litre (meq/L) A weight-per-volume measurement obtained by dividing the concentration expressed in milligrams per litre by the equivalent weight of the substance or ion.

milligrams per litre (mg/L) A weight-per-volume measurement that expresses the concentration of a solute. When specific gravity is unity, a milligrams-per-litre value equals the parts-per-million (ppm) value. When specific gravity is not unity, a milligrams-per-litre value divided by specific gravity of the solution equals the parts-per-million value.

module A membrane element combined with the membrane element housing; a pressure vessel containing one or more membrane elements.

NaHMP Sodium hexametaphosphate.

nanofiltration (NF) A crossflow membrane separation process that removes particles in the 300 to 1,000 molecular weight range, selected salts, and most organics.

NOM Natural organic matter.

nonalkaline scale *See* hard scale.

noncarbonate hardness Hardness caused by chlorides, sulfates, and nitrates of calcium and magnesium. Evaporation of waters containing these ions makes the water highly corrosive.

ntu Nephelometric turbidity unit.

O&M Operations and maintenance.

OEM Original equipment manufacturer.

operating pressure The pressure at which feedwater enters a device.

osmosis The naturally occurring transport of water through a membrane from a solution of low salt content to a solution of high salt content in order to equalize salt concentrations.

osmotic pressure A measurement of the potential energy difference between solutions on either side of a semipermeable membrane. The applied pressure must first overcome the osmotic pressure in the chemical solution for satisfactory reverse osmosis equipment performance.

parts per billion (ppb) A measure of proportion by weight, reflecting the number of unit weights of solute per billion unit weights of solution.

parts per million (ppm) A measure of proportion by weight, reflecting the number of unit weights of solute per million unit weights of solution (approximately equal to milligrams per litre in dilute solutions).

permeability The capacity of a membrane to allow water or solutes to pass through.

permeate channel spacer *See* product channel spacer.

permeate stream A membrane output stream that typically contains a desirable quantity of constituents and is to be used as a product.

permeator A reverse osmosis production unit consisting of the membranes and pressure vessel.

plant capacity A plant's volume production of permeate per unit time.

pore An opening in a membrane or filter matrix.

porosity The proportion, usually stated as a percentage, of the total volume of material that consists of pore space or voids.

posttreatment One or more processes that may be used on the product water, such as chlorination or neutralization. Posttreatment of concentrate, such as pH adjustment, may also be required before disposal.

precipitate A substance separated from a solution by chemical or physical change as an insoluble amorphous or crystalline solid.

pressure filtration Filtration aided by imposing a pressure drop across an enclosed filter vessel.

pressure vessel Several membrane elements in series contained in a single tube.

pretreatment The processes such as chlorination, clarification, coagulation, acidification, and degasification that may be used on the feedwater to a membrane system to minimize algae growth, scaling, and corrosion.

product channel spacer The knit fabric through which permeate water flows after it passes through a flat sheet membrane.

raw water *See* source water.

product stream *See* permeate stream.

recovery The ratio of the permeate flow to the feed flow, generally expressed as a percentage.

reverse osmosis (RO) The transport of water from a solution having a high salt concentration to one having a low salt concentration through a membrane by applying pressure to the solution having a high salt concentration. RO removes ionized salts, colloids, and organics down to 150 molecular weight. It may also be called hyperfiltration.

RIB Rapid infiltration basin.

scaling The precipitation of inorganic salts on the feed side of a membrane.

SDI Silt Density index.

SDWA Safe Drinking Water Act.

solids rejection The percentage of mass removed from the feedwater.

semipermeable membrane A membrane that is permeable only by certain molecules or ions. For example, reverse osmosis membranes will allow water but not salt to pass.

sequestering agent An agent added to feedwater to extend the limits of saturation of scaling substances. The agent ties up and inactivates certain metal ions.

SHMP Sodium hexametaphosphate.

SMBS Sodium metabisulfite.

SOC Synthetic organic chemical.

softener Water treatment equipment that uses a sodium-based ion exchange resin principally to remove calcium and magnesium cations.

soft scale Scale that dissolves under acidic conditions. It is mainly composed of calcium carbonate and magnesium hydroxide.

solubility A measure of the maximum amount of a certain substance that can dissolve in a given amount of water at a given temperature.

solute Matter dissolved in a solvent, typically water.

solution A homogenous mixture of substances in which the molecules of the solute are uniformly distributed among the molecules of the solvent.

solvent A liquid medium that carries dissolved substances, or solutes, typically water.

source water Water that has not been treated on-site, including untreated water from wells, surface sources, the sea, or public water supplies.

spiral-wound cartridge The heart of a spiral-wound desalination device, consisting of the product tube, membrane leaves, feed channel spacers, antitelescoping devices, and brine seal.

spiral-wound membrane *See* flat sheet membrane.

stage Pressure vessels installed in parallel. For example, it is common for a membrane array to have three stages, with four pressure vessels in the first stage, two in the second, and one in the third.

sterilization Destruction or removal of all viable organisms.

supersaturation A state in which the inorganic salts are in solution at a level such that the respective solubility product is exceeded.

TDS *See* total dissolved solids.

telescoping A movement of the outer layers of a spiral-wound cartridge in the direction of the feed flow, caused by excessive flow through the feed channel spacer.

THM *See* trihalomethane.

thrust collar A plastic cylinder located between the last spiral-wound cartridge and end plate to support the last cartridge in a pressure vessel. It has the same diameter as the inside diameter of the pressure vessel.

TOC Total organic carbon.

total dissolved solids The sum of all dissolved solids, volatile and nonvolatile.

TOX Total organic halides.

TOXFP Total organic halide formation potential.

train A membrane arrangement of multiple stages in series where the concentrate is typically used as feed to the subsequent stage.

trihalomethane (THM) Any of several derivatives of methane, CH_4, in which three halogen atoms (chlorine, bromine, or iodine) are substituted for three hydrogen atoms.

turbidity Any undissolved materials in water, such as finely divided particles of sand or clay, reducing the penetration of light and causing the water to appear cloudy.

ultrafiltration (UF) A process using a semipermeable membrane under a hydraulic pressure gradient to separate components in a solution. The membrane pores allow passage of the solvent but will retain nonionic components primarily on the basis of physical size.

USEPA United States Environmental Protection Agency.

water transport The passage of water through a membrane. Water transport is desirable in reverse osmosis and nanofiltration and undesirable in electrodialysis.

Index

NOTE: An *f.* following a page number refers to a figure. A *t.* refers to a table.

This page intentionally blank.